The Drone War

How Unmanned Systems Transformed the Russia-Ukraine Conflict

J.F. Morgenthau

Copyright and Legal Disclaimer

DISCLAIMER

This book is a work of analysis and commentary based on publicly available information, open-source intelligence, media reports, official statements, and documented events related to the Russia-Ukraine conflict from 2022 to 2025. While every effort has been made to ensure accuracy, the author and publisher make no representations or warranties regarding the completeness, accuracy, or reliability of the information contained herein.

The views, opinions, and conclusions expressed in this book are those of the author and do not necessarily reflect the official policy or position of any government, military organization, defense contractor, or other entity. This book should not be construed as official doctrine, policy recommendation, or authoritative military guidance.

Military and Technical Information: Technical specifications, operational details, and tactical information presented in this book are derived from publicly available sources. Sensitive operational security information has been deliberately excluded. Readers should not attempt to implement any tactical or technical procedures described herein without proper authorization, training, and legal authority.

Ongoing Conflict: This book addresses an active military conflict. Circumstances, casualty figures, territorial control, production capabilities, and other details may have changed since publication. The author acknowledges that information about ongoing military operations is inherently incomplete and subject to verification challenges, propaganda, and operational security restrictions.

Ethical Considerations: This book includes discussion of war crimes, civilian casualties, and controversial military operations. The inclusion of such material is for historical documentation and analytical purposes and does not constitute endorsement or justification of illegal actions.

No Professional Advice: This book is not intended to provide military, legal, diplomatic, or policy advice. Readers requiring such guidance should consult with qualified professionals in the relevant fields.

Source Attribution: Where specific sources are cited, the author has made reasonable efforts to verify information through multiple independent sources where possible. However, in the context of active warfare, complete verification is often impossible. Production figures, casualty estimates, and operational claims should be understood as best available information rather than definitive fact.

Limitation of Liability: The author and publisher shall not be liable for any loss of profit or any other commercial damages, including but not limited to special, incidental, consequential, or other damages arising from the use of information contained in this book.

International Readers: Laws regarding discussion of military technology, defense systems, and warfare vary by jurisdiction. Readers are responsible for ensuring their use of this book complies with applicable local laws and regulations.

Living Document Notice: Given the ongoing nature of the conflict and the rapid evolution of drone warfare technology, readers are encouraged to seek updated information from current sources when making decisions based on the content herein.

Table of Contents

Introduction

The sky over Ukraine tells a story that military planners worldwide are only beginning to comprehend. Every night, the distinctive buzzing of Iranian-designed Shahed drones fills the darkness as Russia launches waves of attacks against Ukrainian cities. Every day, Ukrainian FPV pilots guide small quadcopters loaded with explosives toward Russian positions, watching through goggles as their targets approach at speeds exceeding 100 kilometers per hour. Between these attacks and counterattacks, reconnaissance drones by the thousands observe every movement, transmit targeting data, and map the battlefield in real-time. This is modern warfare not the precision-guided, technologically sophisticated combat that Western militaries prepared for, but something rawer, more brutal, and infinitely more prolific.

The numbers alone tell a revolutionary story. In 2021, Ukraine manufactured zero military drones domestically. By October 2025, the country produces 2.5 million annually and is expanding toward 4 million. Russia launches over 5,600 drones in the first nine months of 2024, with attacks intensifying through 2025. A single night in October 2025 saw over 400 Russian drones penetrate Ukrainian airspace. Individual Ukrainian manufacturers produce 40,000 drones monthly. The cumulative drone employment since February 2022 measures in the millions, with both sides consuming and replacing systems at rates that would have seemed fantastical to military planners just five years ago.

These numbers represent more than statistics, they mark fundamental transformation in warfare's character. The conflict that began with expectations of swift conventional victory has evolved into something unprecedented: the world's first large-scale drone war, where unmanned systems dominate operations, determine tactical outcomes, and shape strategic calculations. What started with improvised modifications of commercial quadcopters has become industrialized warfare featuring millions of purpose-built military drones, sophisticated manufacturing ecosystems, and operational concepts developed not in peacetime study but through continuous combat.

This book chronicles that transformation, examining how Ukraine and Russia stumbled into drone warfare's future, how both sides scaled production from nothing to millions annually, and what their experience reveals about conflicts yet to come. The story encompasses technological innovation, industrial mobilization, tactical evolution, and human adaptation to warfare conducted increasingly through remote platforms. It is ultimately a story about how military revolution happens not through careful planning or deliberate strategy, but through desperate improvisation under existential pressure that discovers possibilities no peacetime planner would have predicted.

Why This War Matters

Russia's invasion of Ukraine in February 2022 was not supposed to inaugurate a military revolution. It was supposed to be a swift fait accompli, another demonstration of hard power overwhelming weaker neighbors, concluded before the international community could respond effectively. Instead, it became the laboratory where twenty-first century warfare was forged, where technologies long discussed theoretically were proven operationally, and where assumptions underlying decades of military planning were tested and often found wanting.

The war matters because it provides empirical answers to questions that previous conflicts left unresolved. Can cheap, numerous systems defeat expensive, sophisticated platforms? Yes, demonstrably and repeatedly. Can civilian volunteers contribute meaningfully to advanced military operations? Yes, Ukrainian volunteer networks have procured and produced capabilities that formal government procurement could not match. Can industrial capacity be mobilized rapidly from essentially nothing? Yes, Ukraine's drone industry grew from non-existence to world-leading scale in under four years. Can attrition warfare return in an age of precision weapons? Yes, the grinding materiel consumption and replacement that characterized twentieth century total wars dominates this conflict despite all the technological advancement since.

These answers matter because they apply far beyond Ukraine's borders. Every military worldwide is studying Ukrainian experience, extracting lessons, and adapting force structures and procurement priorities accordingly. NATO is reorganizing around drone warfare, acquiring millions of systems and restructuring doctrine comprehensively. China is accelerating drone programs while analyzing implications for potential Taiwan scenarios. Regional powers are establishing drone production capabilities. Non-state actors are incorporating Ukrainian innovations into their own operations. The proliferation is global, rapid, and irreversible.

The war also matters because it demonstrates warfare's democratization through technology. When FPV drones costing hundreds of dollars can destroy tanks worth millions, when garage workshops can produce militarily significant capabilities, when teenagers trained in weeks can conduct strikes as effectively as professional soldiers, the barriers to military effectiveness lower dramatically. This democratization has profound implications for terrorism, insurgency, regional conflict, and the global security order. Ukraine has opened Pandora's box, revealing that sophisticated military capabilities once requiring nation-state resources are now accessible to any group with modest funding and technical expertise.

A Note on Sources and Perspective

This book draws on multiple sources: open-source intelligence from Ukrainian and Russian social media, official military statements and press releases, journalistic coverage from multiple nations, academic analysis from defense institutions, economic data on procurement and production, and extensive documentation of the conflict available through various platforms. Where possible, information has been corroborated across multiple independent sources. Numbers regarding production rates, attack volumes, and operational details come from official Ukrainian government statements, defense ministry releases, and corroborated reporting from established media outlets.

The perspective is necessarily partial. Complete information about ongoing military operations is unavailable and much remains classified or deliberately obscured by operational security. Russian information is particularly difficult to verify given that nation's information controls and deliberate disinformation campaigns. Ukrainian information, while generally more transparent, reflects natural bias toward presenting military operations favorably and concealing vulnerabilities. This book attempts to navigate these limitations by focusing on verifiable facts, acknowledging uncertainty where it exists, and drawing conclusions only where evidence is substantial.

The analysis is unapologetically focused on military effectiveness and strategic implications rather than political judgments about the war's justification. Russia's invasion represents clear aggression under international law and moral condemnation is warranted, but this book's purpose is examining how warfare has evolved rather than prosecuting legal or ethical cases. Where war crimes are discussed, particularly regarding drone-delivered attacks on civilians, the documentation is presented factually, allowing readers to draw appropriate conclusions about culpability and accountability.

Why You Should Read This Book

If you are a military professional, policymaker, defense analyst, or student of warfare, this book provides a comprehensive examination of the most significant military transformation since World War II. The lessons from Ukraine will shape defense planning, procurement priorities, and operational concepts for decades. Understanding what happened, why it happened, and what it means for future conflict is essential for anyone involved in national security.

If you are simply a concerned citizen trying to understand the world's trajectory, this book illuminates one of the defining developments of our era. Drone warfare is not a distant future or science fiction but a present reality that will affect global security, terrorism threats, regional stability, and potentially your own safety in ways both direct and indirect. Ukraine's experience provides a window into this transformed world and preview of conflicts yet to come.

If you care about Ukraine specifically, this book documents a dimension of Ukrainian resistance that deserves recognition alongside military valor and civilian resilience. Ukraine's innovation under pressure, its industrial mobilization from nothing to world-leading scale, and its contribution to global understanding of warfare represent achievements as significant as battlefield victories. Ukraine has become an inadvertent teacher to the world, and its lessons are written in the sky over its cities and battlefields.

The drone age is here. It arrived not gradually but explosively, compressed into three and a half years of high-intensity conflict that demonstrated possibilities and revealed vulnerabilities that peacetime speculation could never have uncovered. What Ukraine and Russia have learned through terrible cost, the world must now understand and adapt to. This book provides that understanding, drawn from the most comprehensive case study of modern warfare available, documented as it unfolded and analyzed with perspective that only distance from immediate combat allows.

Welcome to the future of warfare. It is not what anyone expected, not what anyone wanted, but it is undeniably what warfare has become. The sky over Ukraine tells this story every day and night, written in the contrails of thousands of drones, in the explosions of strikes and interceptions, and in the transformed tactical and strategic landscape that emerges from this revolution. This book translates that sky-written story into understanding, analysis, and lessons that no one interested in warfare, security, or humanity's future can afford to ignore.

PART I: THE AWAKENING (February 2022 - December 2022)

Chapter 1

The War Nobody Expected

February 24, 2022

The first explosions shook Kyiv at 5:00 AM. For weeks, Western intelligence agencies had warned that Russian President Vladimir Putin was preparing to invade Ukraine, but even as satellite images showed more than 150,000 troops massed along the border, many refused to believe it would actually happen. The scale of what Putin had unleashed, a full-scale invasion of Europe's second-largest country defied contemporary expectations of how modern wars were fought.

Within hours of the initial missile strikes, Russian forces poured across Ukraine's borders from three directions: from Belarus in the north, driving toward the capital Kyiv; from Russia's western regions in the northeast and east; and from occupied Crimea in the south. The scope was breathtaking and terrifying. This wasn't a limited operation or a localized conflict. This was the largest military invasion in Europe since World War II.

Putin's calculus appeared straightforward, even if grotesquely miscalculated. Intelligence assessments would later reveal that the Kremlin expected to control Ukraine within ten days. The plan relied on speed, shock, and the assumption that Ukrainian resistance would crumble in the face of overwhelming force. Russian troops carried dress uniforms in their vehicles, expecting to parade through Kyiv as liberators. Special operations forces were tasked with seizing government buildings and capturing or killing President Volodymyr Zelensky. Once the government fell, Moscow would install a puppet regime and the war would be over before the West could mount a meaningful response.

It was a fantasy built on faulty intelligence, wishful thinking, and a fundamental misunderstanding of Ukrainian resolve. But in those first chaotic hours and days, no one knew for certain what would happen. The world watched and waited, expecting the worst.

The Ghost of Wars Past

Military analysts worldwide expected to witness a replay of recent conflicts, where overwhelming technological superiority determined outcomes swiftly and decisively. The United States' Operation Desert Storm in 1991 and Operation Iraqi Freedom in 2003 had demonstrated what modern combined arms warfare looked like when executed by a superpower: precision strikes, total air superiority, coordinated ground offensives, and the systematic destruction of enemy command and control. Israel's Six-Day War in 1967 provided an even more dramatic template, rapid and decisive victory through technological and tactical superiority.

Russia, possessing the world's second-most powerful military on paper, should have been able to replicate something similar. They had the jets, the tanks, the missiles, the manpower. Ukrainian forces, while determined and increasingly well-trained since the 2014 Russian incursion into Donbas, were outnumbered and outgunned. Common military wisdom suggested that motivation and courage, while important, couldn't overcome such disparities in equipment and numbers.

Yet within days, the carefully laid Russian battle plans began to unravel. The lightning strike on Kyiv stalled. The famous forty-mile convoy of Russian vehicles approaching the capital ground to a halt, harassed by Ukrainian forces and crippled by logistical failures. Ukrainian defenders didn't collapse; they fought back with unexpected ferocity and, critically, with unexpected effectiveness.

What military observers were witnessing, though many didn't realize it yet, was the beginning of a new type of warfare. The conflict that unfolded in Ukraine wouldn't resemble Desert Storm or Iraqi Freedom. It would look more like the First World War than the conflicts of the 21st century, a grinding attritional struggle fought across vast fronts with artillery, trenches, and enormous casualties on both sides. But there would be one crucial difference that would define this war and transform military thinking worldwide: the pervasive, inescapable presence of drones.

Eyes in the Sky

From the very first days of the invasion, unmanned aerial vehicles played a role that few had anticipated. Ukraine possessed a modest fleet of Turkish-made Bayraktar TB2 drones, medium-altitude, long-endurance unmanned combat aerial vehicles that had proven themselves in previous conflicts in Libya, Syria, and Azerbaijan's war with Armenia in 2020. At the start of the full-scale invasion, Ukraine had approximately twenty operational TB2s split between the Air Force and Navy. By early March 2022, they had received an additional sixteen units, bolstering their aerial reconnaissance and strike capabilities.

The TB2s became Ukraine's unlikely heroes in the war's opening phase. While Ukrainian pilots struggled to contest Russian air superiority with their limited fleet of aging fighters, the Bayraktars hunted Russian columns with devastating effectiveness. Videos began circulating on social media showing grainy infrared footage of Russian trucks, tanks, and artillery pieces being destroyed by precision strikes. Each video was accompanied by the now-iconic theme song "Bayraktar," which became an anthem of resistance.

The drones' early success stemmed not from any overwhelming technological advantage, the TB2 is relatively simple compared to advanced Western systems but from Russian disorganization and the specific conditions of the conflict's opening phase. Russian forces had been slow to establish proper air defense networks as they advanced. Their columns stretched for miles along exposed roads, presenting perfect targets for aerial attack. Ukrainian operators exploited these vulnerabilities ruthlessly.

One of the TB2's most celebrated moments came in April 2022, when it played a reconnaissance role in the sinking of the Russian Black Sea Fleet flagship, the Moskva. The drone reportedly provided targeting data and distracted Russian air defenses, allowing Ukrainian Neptune anti-ship missiles to strike home. The loss of the Moskva, the first sinking of a fleet flagship in combat since World War II dealt a humiliating blow to Russian prestige and demonstrated how drones could enable conventional weapons to achieve disproportionate effects.

On Snake Island, a small rocky outpost in the Black Sea that had become a symbol of Ukrainian defiance, TB2s proved devastatingly effective at preventing Russian attempts to establish a permanent presence. The drones loitered overhead, striking Russian helicopters attempting to deliver supplies and destroying air defense systems as soon as they were set up. Buildings were razed, making the barren rock even more bare. The Russians eventually abandoned the island, unable to sustain their position under constant aerial surveillance and attack.

The Democratization of Reconnaissance

While the Bayraktars captured headlines and imaginations, a quieter revolution was occurring across Ukrainian front lines. Soldiers were pulling consumer drones, primarily DJI Mavics and similar commercial quadcopters out of their personal bags and launching them to scout enemy positions. These were the same drones that hobbyists used to film landscapes and real estate agents employed for aerial photography. Now they were being pressed into military service out of necessity.

The transformation was organic and unplanned. Some units had acquired commercial drones before the war as part of their training. Others bought them with personal funds or received them from volunteer organizations that sprang up across Ukraine and the international community. Within weeks, thousands of these small, inexpensive drones were operating along the front lines, providing Ukrainian forces with a capability that had been the exclusive province of militaries with satellite networks and sophisticated reconnaissance aircraft: real-time aerial surveillance.

For artillery units, the impact was immediate and profound. Traditional artillery operation relied on forward observers who had to position themselves to see the target and the impact of shells, a dangerous and imprecise art. With a drone overhead, artillery crews could watch their rounds land in real-time, making adjustments instantaneously. What might have taken dozens of shells to accomplish through traditional methods could now be done with a handful. Ammunition, always precious in war, became more effective by an order of magnitude.

Infantry commanders gained the ability to see over hills, around buildings, and into enemy positions before committing their troops. Ambushes could be spotted. Enemy movements could be tracked. The fog of war, which had obscured battlefields since time immemorial, began to lift. Not completely, never completely but enough to give Ukrainian defenders a crucial edge against a more numerous enemy.

The Russians had drones too, of course, and in greater numbers initially. But Ukraine's forces adapted faster, developed tactics more quickly, and integrated commercial drones into their operations with a flexibility that the more rigid Russian command structure struggled to match. Ukrainian soldiers, many of whom had grown up with smartphones and video games, proved natural drone operators. They shared footage, tactics, and modifications through Telegram channels and social media, creating a distributed network of innovation that no central command could have organized.

The Great Surprise

As winter gave way to spring in 2022, it became increasingly clear that Russia's invasion had failed in its primary objective. Kyiv remained in Ukrainian hands. Zelensky was not dead or in exile but instead was rallying his nation and the world from his offices in the capital. Russian forces had withdrawn from the north, leaving behind evidence of atrocities in places like Bucha that hardened Ukrainian resolve and international support.

The war had settled into something resembling the static warfare of the Western Front in 1914-1918, with elaborately fortified positions, artillery duels, and grinding battles over small parcels of territory. Russia concentrated its efforts on the eastern Donbas region, seeking to complete the conquest of Luhansk and Donetsk oblasts. Progress was measured in hundreds of meters, paid for in blood and steel.

But above the trenches, something new was emerging. The number of drones operating on both sides continued to multiply. Ukrainian volunteers abroad launched crowdfunding campaigns to buy more commercial drones. Defense manufacturers ramped up production. Soldiers on leave would bring drones back to the front when they returned from rotation. Russian forces deployed their own growing fleets of reconnaissance and strike drones.

What had begun as an improvised use of commercial technology was evolving into something more systematic and more deadly. Engineers began modifying consumer drones to carry small munitions, grenades initially and then custom-made explosives. The drones that had started the war as scouts were becoming hunters. First-person view racing drones, normally used for high-speed competitions where pilots wore goggles to see from the drone's perspective, were being weaponized. Their speed and agility made them difficult to shoot down, and their low cost meant they could be risked on dangerous missions.

By summer 2022, the transformation was well underway, though few yet grasped its full implications. The TB2s, which had seemed so dominant in the war's opening months, were appearing less frequently. Russian air defenses had been established, and the Turkish-made drones proved vulnerable to sophisticated anti-aircraft systems. At least twenty-six had been destroyed by March 2022, according to verified visual evidence, with actual losses likely significantly higher. The era of the Bayraktar as Ukraine's primary aerial weapon was already waning.

But that didn't matter, because something else was rising to take its place. Something cheaper, more numerous, and more adaptable. The age of mass drone warfare was beginning, though no one had a name for it yet. Military doctrine written for an earlier era was becoming obsolete in real-time. Officers trained on conventional warfare were learning new tactics through trial and error. Soldiers who had never considered careers in aviation were becoming skilled drone pilots.

The Shape of Things to Come

By December 2022, as the first year of the war drew to a close, the front lines had largely stabilized. Russia controlled approximately 18 percent of Ukrainian territory, including Crimea which it had seized in 2014. Both sides were digging in for what was clearly going to be a long war. International attention, which had been fixated on Ukraine in the early months, was beginning to drift. The war had become normalized, a steady drumbeat of bad news that no longer dominated headlines unless something particularly dramatic occurred.

What wasn't yet apparent to casual observers but was becoming crystal clear to those fighting and dying in the fields of eastern Ukraine was that this war was writing the future of combat. The drones that had been improvised solutions in February had become essential tools by December. Units that had operated with a handful of commercial quadcopters now had dozens. Specialized drone operators were emerging as a distinct military specialty. Competition for electronic warfare specialists, soldiers who could jam enemy drones or protect friendly ones was becoming fierce.

On both sides, military thinkers were beginning to understand that they were participants in something unprecedented. Ukraine, the invaded nation fighting for its survival, had strong incentives to innovate. They couldn't match Russia's industrial capacity for producing traditional weapons, but drones were different. Drones could be built in workshops and garages. Drone operators could be trained in weeks rather than years. Drone tactics could be developed by the soldiers themselves, without waiting for commands from distant generals.

Russia, despite its initial setbacks, possessed enormous resources and was beginning to mobilize its military-industrial complex for extended war. The Kremlin was also looking abroad for solutions, particularly to Iran, which had developed its own domestic drone industry under decades of international sanctions. Russian and Iranian officials were in discussions about technology transfers that would reshape the war in the coming year.

Neither side yet understood the full scope of what was coming. The thousands of drones operating in late 2022 would become tens of thousands in 2023, then hundreds of thousands in 2024, and eventually millions. The small modifications and improvisations of the war's first year would evolve into sophisticated manufacturing ecosystems. The tactical innovations being developed in Ukrainian trenches would eventually influence military thinking in Beijing, Washington, and every capital in between.

The war that nobody expected was teaching lessons that nobody wanted to learn. It was demonstrating that the future of warfare wouldn't be dominated by stealth fighters and aircraft carriers, by tanks and artillery, or even by precision-guided missiles. Those systems would remain important, but the future belonged to cheap, numerous, expendable platforms that could be produced faster than they could be destroyed and evolved faster than defenses could be developed.

The Russians had expected to win in ten days. The Ukrainians had expected to fight desperately for survival against impossible odds. The world had expected to witness either a quick Russian victory or a protracted guerrilla war. Instead, everyone was getting something different: a glimpse of warfare's future, where the sky belonged not to billion-dollar fighter jets but to plastic and foam devices that cost less than a smartphone, where victory wasn't determined by who had the most advanced technology but by who could produce, deploy, and adapt the fastest.

As 2022 ended and the war ground on into its second winter, Ukrainian and Russian soldiers manning frozen trenches looked up at the same cold sky. Somewhere up there, too small to see and too quiet to hear, drones were watching. They were always watching now. And increasingly, they were killing. The drone war had begun, though its full fury was yet to come. The war nobody expected was becoming the war that would change everything.

Chapter 2

Consumer Drones Go to War

A Toy Becomes a Weapon

In March 2022, a Ukrainian artillery unit stationed near Brovary, just east of Kyiv, faced a problem that had plagued gunners since the invention of artillery itself: they couldn't see their target. Russian armored vehicles were advancing along a road several kilometers away, beyond the range of direct observation. In previous wars, the solution would have involved forward observers risking their lives to get close enough to spot targets and call in fire. Radio communications would crackle back and forth, adjustments made by educated guesswork, with dozens of shells expended to achieve hits.

But this wasn't a previous war. A young soldier pulled a compact folding drone from his backpack, a DJI Mavic 3, the kind of quadcopter that photography enthusiasts and real estate agents had been using for years. He launched it with a few button presses on a controller connected to his smartphone. Within minutes, the drone was hovering two hundred meters above the Russian column, its camera transmitting a crystal-clear video feed. The artillery commander watched on a tablet as the first shell landed fifty meters short. "Up fifty," he said calmly. The next round landed thirty meters long. "Down twenty." The third shell struck directly on target, and the rest of the battery joined in with devastating accuracy.

What had once required extensive training, dangerous reconnaissance, and considerable luck now took minutes and a device that cost less than three thousand dollars. The Russian vehicles, unable to see or hear the tiny drone circling overhead, became targets in a shooting gallery. By the time they realized what was happening and attempted to retreat, half the column had been destroyed.

This scene, repeated thousands of times across Ukraine's battlefields in 2022, represented a revolution in warfare that few military theorists had anticipated. Small off-the-shelf DJI Mavic quadcopters, priced around

$1,500 to $3,000, became widely used for frontline reconnaissance and artillery correction. The democratization of aerial surveillance, once the exclusive province of nation-states with satellite networks and sophisticated military aircraft, was now available to any soldier with a credit card and an internet connection.

The DJI Phenomenon

The drones transforming warfare in Ukraine came primarily from a single manufacturer: DJI, a Chinese company that had dominated the consumer and commercial drone market since the mid-2010s. DJI's Mavic series, particularly the Mavic 3, combined remarkable capabilities with user-friendly operation. They could fly for up to forty minutes on a single battery, reach altitudes of several kilometers, and transmit high-definition video over distances of ten kilometers or more. Built-in GPS provided stable hovering and autonomous flight capabilities. Most importantly, they were designed to be operated by amateurs, no specialized training required.

In Ukraine, Mavics became the de facto standard for small unit reconnaissance and artillery spotting, with 'Mavik' used generically as a term for consumer quadcopters on the battlefield. The term became so ubiquitous that soldiers would refer to any small reconnaissance drone as a "mavic," regardless of manufacturer, much as Americans once called all tissues "Kleenex" and all petroleum jelly "Vaseline."

The reasons for DJI's military adoption went beyond mere availability. The drones were reliable, with millions of units produced and tested in civilian applications worldwide. Parts were easy to source, and repairs could be conducted in the field by soldiers with basic technical skills. The control systems were intuitive, if you could operate a smartphone, you could fly a Mavic. Most crucially, they were mass-produced commercial products, meaning they were relatively inexpensive and could be acquired in large quantities without going through traditional military procurement channels.

Ukrainian units quickly discovered that the drones' commercial features translated remarkably well to military applications. The thermal cameras offered on higher-end models, designed for industrial inspections and search-and-rescue operations, proved ideal for detecting enemy soldiers and vehicles at night. Zoom lenses allowed operators to identify targets from safe distances. The ability to pre-program flight paths meant drones could conduct reconnaissance missions autonomously, reducing operator exposure to enemy fire.

Russian forces used DJI drones as well, often in greater numbers initially due to their larger military budget and easier access to international markets. But Ukraine's forces adapted faster, integrated the technology more effectively, and developed innovative tactics that would define the drone war's early phase.

The Volunteers' War

As the invasion's first weeks unfolded, a remarkable grassroots movement emerged across Ukraine and the international community. Volunteer organizations sprang up with a single purpose: getting drones to Ukrainian soldiers. What began as ad-hoc efforts by individuals, techies buying drones with their own money and driving them to the front rapidly evolved into sophisticated international networks capable of raising millions of dollars and delivering thousands of drones.

Serhiy Prytula, a politician and army volunteer, and Serhii Sternenko, an activist, established a crowdfunding site that raised $9.6 million in just 24 hours. The speed and scale of the response stunned even the organizers. Ukrainians inside the country and members of the diaspora worldwide contributed. International supporters, moved by Ukraine's resistance, donated whatever they could afford. The funds poured in faster than they could be spent, and more campaigns launched to meet the insatiable demand.

These weren't government programs or traditional military procurement. This was warfare crowdfunded by ordinary people who understood viscerally that technology could save lives. A drone spotting an ambush meant Ukrainian soldiers came home. A drone guiding artillery accurately meant fewer shells wasted and more Russian vehicles destroyed. Every donation, every device delivered, tilted the odds slightly more in Ukraine's favor.

The volunteer networks developed their own logistics systems, often more efficient than official channels. Drones purchased in Poland, Germany, or the United States would be transported to Ukraine by volunteers, sorted at distribution hubs, and delivered directly to units that requested them. Some volunteers specialized in procurement, using their connections in the tech industry to source drones at bulk discounts. Others focused on logistics, arranging transport across borders. Still others provided training, teaching soldiers how to operate and maintain their new equipment.

Social media became the coordination backbone for these efforts. Telegram channels connected donors with recipients. Twitter campaigns went viral, raising awareness and funds. YouTube videos showed drones in action, demonstrating to potential donors exactly how their money would be used. The transparency of the process, seeing the actual drones, meeting the actual soldiers who would use them created a direct connection between civilian supporters and military outcomes rarely seen in modern warfare.

Innovation on the Fly

Ukrainian soldiers, many of whom had grown up with smartphones and video games, proved remarkably adept at finding creative applications for their new tools. The drones arrived as commercial products, but they didn't stay that way for long. Within weeks, Ukrainian units were modifying them, pushing their capabilities far beyond what the manufacturers intended.

Drones were modified on an industrial scale with 'hacks' to prevent the drone from broadcasting its location, a critical adaptation since Russian forces could use the drones' radio signals to track them back to their operators. Ukrainian technicians disabled or modified the electronic identification systems that DJI had built into their drones to comply with aviation regulations. They experimented with different antennas to extend range and improve signal reliability in the face of Russian jamming efforts.

Some modifications were purely practical. Soldiers painted drones in camouflage patterns to reduce visual detection. They developed quick-release mounts for batteries, allowing faster swaps and more flight time per day. They created custom carrying cases designed for military vehicles and field conditions, protecting the delicate electronics from the rough environment of the front lines.

Other innovations were more aggressive. Units began experimenting with weaponizing their reconnaissance drones. The early attempts were crude, a hand grenade attached to the drone with tape, released by a servo mechanism controlled from the ground. But they worked. Videos appeared on Telegram and Twitter showing small commercial drones dropping grenades through the open hatches of Russian tanks, through the sunroofs of vehicles, and into trenches. The psychological impact was immense. Russian soldiers learned to fear the buzzing of small drones overhead, never knowing if they were being watched or about to be attacked.

The R18 octocopter, a heavy-lift platform designed for professional cinematography, found new purpose in Ukraine. The R18 could fly without lights in the dark for total visual stealth, while using thermal-imaging cameras to spot enemy troops and vehicles, with videos showing Russian tanks being wiped out by cheap, armor-piercing grenades fitted with 3D-printed tail fins. Ukrainian engineers and soldiers, working together, had transformed a tool designed for filmmakers into a deadly nighttime hunter.

The Tactical Revolution

The impact of commercial drones on battlefield tactics in 2022 cannot be overstated. Artillery units saw their effectiveness increase by an order of magnitude. What previously required hours of preparation and dozens of shells could now be accomplished in minutes with a handful of rounds. Ammunition, always a precious commodity in war, went further. Supply chains came under less strain. Logistics officers could deliver more ammunition to more units with the same resources.

Infantry operations changed fundamentally. Commanders could now scout enemy positions without risking soldiers' lives. Ambushes that would have succeeded through concealment could be spotted from above. Ukrainian forces defending against Russian attacks had a God's-eye view of approaching enemy formations, allowing them to position their limited forces optimally and strike at the most vulnerable moments.

The advantage wasn't total, Russian forces had drones too, and in some areas deployed them more heavily. But Ukraine's more flexible command structure, younger and more tech-savvy soldiers, and the urgency of defending their homeland created an innovation culture that Russia's more rigid military struggled to match. Ukrainian soldiers shared discoveries through social media and messaging apps. A tactic developed by one unit would spread across the entire front within days. Russian forces, operating under stricter communications discipline and with less trust in their soldiers' initiative, couldn't match this pace of adaptation.

The effect extended beyond immediate tactical advantages. Drone footage changed how the world saw the war. Every destroyed Russian vehicle, every failed assault, every Ukrainian success was documented from above and shared globally. This transparency served multiple purposes: it maintained international support for Ukraine, it boosted Ukrainian morale by demonstrating their forces' effectiveness, and it degraded Russian morale by providing incontrovertible evidence of their failures.

Ukrainian psychological operations officers quickly grasped the power of aerial footage. Drone videos of destroyed Russian convoys were edited with dramatic music and released online, garnering millions of views. The footage was visceral and immediate in ways that traditional war reporting could never match. Viewers could see exactly what was happening on the battlefield, from the perspective of the Ukrainian soldiers winning small victories every day.

The Dark Side of Democratization

But the democratization of aerial surveillance brought challenges as well. The same transparency that aided Ukrainian operations also meant Russian forces could watch them. Every Ukrainian position visible to Ukrainian drones was potentially visible to Russian drones. Every movement that could be tracked from above could be targeted by enemy artillery. The fog of war was lifting for both sides, leading to a situation where maneuver became extremely dangerous and static defense became the default.

By summer 2022, the front lines had largely stabilized into a form of warfare that would have been familiar to soldiers from World War I. Elaborate trench systems snaked across hundreds of kilometers of

Ukrainian territory. Movement during daylight hours invited drone observation and artillery strikes. Units traveled under cover of darkness or bad weather, using smoke and camouflage to mask their activities from the omnipresent eyes overhead.

This mutual observation created a grim calculus. Both sides could see each other constantly, making surprise attacks nearly impossible. Breakthroughs that required concentrated force became death traps, massing troops for an assault meant presenting a target-rich environment to enemy drones and artillery. The war settled into a grinding attritional struggle where territory changed hands slowly, measured in hundreds of meters rather than kilometers, each advance paid for in blood.

The electronic warfare dimension escalated rapidly. Russian forces deployed increasingly sophisticated jamming systems to blind or disable Ukrainian drones. Ukrainian forces countered with frequency-hopping radios, stronger antennas, and autonomous flight modes that allowed drones to complete missions even when communications were disrupted. The electromagnetic spectrum became a battlefield as important as the physical terrain, with both sides constantly adapting to the other's countermeasures.

Civilian casualties rose as both sides struggled to distinguish between military and non-military targets from altitude. A truck carrying supplies for soldiers or a truck carrying food for civilians often looked identical from above. Buildings housing military communications equipment or buildings housing families could be indistinguishable without ground intelligence. While both sides claimed to take precautions, the reality of war conducted through camera feeds inevitably led to tragic mistakes.

The Supply Chain War

As demand for drones skyrocketed through 2022, supply became a critical issue. DJI, initially willing to sell to both sides, faced increasing international pressure to limit sales to Russia. The company's position became increasingly untenable as evidence emerged of their drones being used in war crimes. In April 2022, DJI announced it would suspend operations in both Russia and Ukraine, though enforcement proved difficult given the thriving gray market.

DJI Mavics, retailing for about $2,000, transitioned from amateur photography tools to critical military assets, with their role expanding early in the 2022 Russia-Ukraine war. The suspension of official sales didn't stop the flow of drones to the battlefield, it just made them more expensive and harder to acquire. Volunteers found ways to purchase drones in third countries and transport them to Ukraine. Units began hoarding spare parts and batteries, knowing resupply would become more difficult.

Russian forces, initially advantaged by easier access to Chinese markets, found their drone supplies constrained as Western sanctions made international trade more difficult. The Chinese government, walking a diplomatic tightrope between Russia and the West, tacitly discouraged companies from supplying either side while doing little to enforce these preferences. DJI drones continued flowing to both armies through complex supply chains that obscured their ultimate destinations.

Ukrainian officials recognized the vulnerability of depending on foreign manufacturers for critical military equipment. Planning began for domestic drone production, though in 2022 the focus remained on acquiring and deploying commercial systems as quickly as possible. The seeds of Ukraine's future drone industry were being planted, but they wouldn't bear fruit until the following year.

The Learning Curve

By December 2022, as the first year of full-scale war drew to a close, the role of commercial drones in the conflict had evolved far beyond what anyone anticipated in February. What began as improvised use of civilian technology had become systematic doctrine. Every Ukrainian brigade had drone operators. Every artillery battery had spotters equipped with Mavics. Every infantry company had soldiers trained in drone reconnaissance.

Training programs that didn't exist in February were now churning out qualified operators by the hundreds. Soldiers rotated through courses where they learned not just how to fly drones, but how to use them effectively for reconnaissance, how to coordinate with artillery, how to maintain and repair them under field conditions. The skills that had been the province of hobbyists and technicians were now essential military competencies.

The Ukrainians discovered that drone operations required more than just technical skill. Good operators needed tactical understanding, knowing what to look for, recognizing patterns in enemy behavior, anticipating likely actions. They needed patience, spending hours watching enemy positions, waiting for the right moment to strike or the right intelligence to report. They needed communication skills, able to convey what they saw clearly and concisely to commanders who needed to make split-second decisions.

Russian forces were learning too, though more slowly. Their early dismissal of commercial drones as toys gave way to grudging respect and then frantic efforts to match Ukrainian capabilities. Russian units began receiving more drones, their soldiers received better training, and their tactics evolved to account for the new reality of warfare under constant surveillance.

But the fundamental asymmetry remained: Ukraine, fighting for survival on its own territory, adapted faster and innovated more aggressively than Russia's larger but more bureaucratic military machine. The lessons learned through 2022 would set the stage for the even more dramatic developments to come. Commercial drones had proven themselves as military tools. The next phase would be even more revolutionary: the weaponization of those drones on an industrial scale, and the birth of entirely new platforms designed specifically for the unique demands of this new kind of warfare.

Chapter 3

The Shahed Arrives

The Sound of Terror

Ukrainians who lived through the autumn of 2022 remember the sound. It started as a distant buzz, like a swarm of angry bees or a poorly tuned motorcycle engine. As it grew closer, the sound became unmistakable, a distinctive, rasping drone that announced death from above. By the time you heard it clearly, the question was no longer whether an attack was coming, but only where the Iranian-made Shahed drone would strike.

The first confirmed appearance of these weapons came on September 13, 2022, when photographs emerged of the wreckage of a drone marked with the Russian designation "Geranium-2" (Герань-2). But according to Ukrainian artillery commander Rodion Kulahin of the 92nd Brigade and multiple media reports, Shahed drones had actually begun appearing on Ukraine's frontlines as early as August 2022. The initial deployments were tactical, targeting military positions near the front lines. Russian forces were testing the new weapon, learning its capabilities, and refining their tactics.

What Ukrainian defenders saw in those first encounters would fundamentally alter the character of the war. These weren't sophisticated military drones like the Turkish TB2s or American Predators. The Shahed-136, as it was known by its Iranian designation, was something else entirely, a one-way attack unmanned aerial vehicle, a guided missile disguised as a drone, built from surprisingly basic materials but devastating in its simplicity.

The design was almost elegant in its brutality. The Shahed-136 measured 3.5 meters in length with a 2.5-meter wingspan, constructed primarily from foam and plywood with critical components made from metal and plastics. It weighed approximately 200 kilograms and carried a warhead of 36 to 50 kilograms of high explosives, more destructive power than a 155mm artillery shell. Its small piston engine, cheap and easy to produce, gave it a range of up to 2,500 kilometers, though it was typically employed at distances of several hundred kilometers in Ukraine.

Unlike the multimillion-dollar precision weapons that dominated modern arsenals, the Shahed was designed for mass production and expendability. Leaked Iranian documents would later reveal that Iran sold approximately 6,000 Shahed-136s to Russia in 2022 at a unit price of $193,000 per drone, expensive by consumer standards but extraordinarily cheap for a weapon that could strike targets hundreds of kilometers behind enemy lines. The documents also indicated that Russia expected the unit cost to drop to just $48,000 once domestic production began.

Iran's Shadow War

The appearance of Iranian drones in Ukraine marked a significant escalation in Iran's involvement in the conflict and a troubling development for Western observers. For years, Iran had been developing its domestic drone industry under the crushing weight of international sanctions that prevented it from acquiring advanced military technology from abroad. Unable to purchase sophisticated weapons systems, Iranian engineers had been forced to innovate, creating indigenous designs that emphasized practicality and affordability over cutting-edge performance.

The Shahed series represented the culmination of this effort. These weren't attempts to match Western or Russian military drones specification for specification. Instead, they filled a specific niche: long-range strike weapons that could be produced in large quantities, operated with minimal training, and deployed in saturation attacks that would overwhelm enemy defenses through sheer numbers. Iran had tested these concepts in the Middle East, providing Shahed drones to Houthi rebels in Yemen who used them to strike Saudi Arabian targets hundreds of kilometers away.

Now Russia, facing its own isolation from Western military technology and struggling with the attrition of its precision weapons stockpiles, was turning to Iran for solutions. The initial deal involved the transfer of fully assembled drones, but it also included something more valuable: technology transfer and production licensing. Russia wasn't just buying weapons; it was buying the capability to manufacture them domestically, using its own supply chains and resources.

The geopolitical implications were profound. Iran and Russia, both under Western sanctions and both seeking to challenge American global influence, were forging a deeper military partnership. For Ukraine, it meant facing weapons developed and tested in Middle Eastern conflicts, adapted now for European warfare. For the West, it demonstrated how sanctioned nations could support each other, sharing technology and expertise to circumvent international isolation.

The Shahed drones themselves bore the fingerprints of this international collaboration. Inspections of downed drones revealed a troubling supply chain: Western electronics smuggled through third countries, Chinese components acquired through gray market channels, Iranian design and assembly, and Russian operational deployment. Sanctions, it turned out, were less impermeable than their architects hoped.

From Tactical to Strategic

Through September and early October 2022, Russian forces employed the Shaheds primarily as battlefield weapons, targeting Ukrainian military positions, ammunition dumps, and command posts near the front lines. These initial attacks demonstrated the drones' capabilities but also their vulnerabilities. Ukrainian air defenses, adapting to the new threat, began achieving regular intercepts. Mobile anti-aircraft systems and even machine gun fire could bring down the relatively slow-moving drones if they were detected in time.

But Russia's military planners had a different vision for the Shaheds. If Ukrainian air defenses could intercept some percentage of the drones, the solution was simple: launch more drones than the defenses could handle. And if the tactical impact of striking front-line positions was limited, why not target something that would hurt Ukraine more deeply, its civilian infrastructure and its people's will to resist?

The strategy shifted in mid-October 2022. On October 10th, Russia launched massive coordinated attacks against Ukrainian cities, using both conventional cruise missiles and Shahed drones. The targets weren't military bases or troop concentrations. They were power plants, electrical substations, water facilities, and heating infrastructure. As winter approached, Russia was attempting to freeze Ukraine into submission.

The Shahed became a weapon of terror, deliberately designed to target civilian morale. The drones were often launched at night, their distinctive engine sound echoing through darkened cities. Air raid sirens wailed. Families huddled in basements and metro stations. The explosions, when they came, were often secondary to the psychological impact of the hours spent waiting, listening to that terrible buzzing growing closer or fading away.

The attacks followed a brutal calculus. Each Shahed cost Russia between $48,000 and $193,000 to produce or purchase. The damage they caused, a destroyed transformer, a damaged water treatment facility, a residential building collapsed cost millions to repair. But more valuable than the physical destruction was the human cost: millions of Ukrainians unable to sleep, going to work exhausted, living in constant fear. This was asymmetric warfare in its purest form, using cheap weapons to inflict disproportionate suffering.

The Defense Dilemma

Ukrainian air defense forces found themselves facing an impossible equation. The Shaheds were slow, cruising at speeds around 185 kilometers per hour making them theoretically easy targets. But "easy" was relative when you were trying to track and destroy dozens of drones simultaneously, each one potentially carrying death to a power plant or a residential neighborhood.

The cost disparity created a strategic nightmare. When Ukraine used a sophisticated surface-to-air missile worth hundreds of thousands or even millions of dollars to shoot down a Shahed worth perhaps $50,000, Russia was winning the exchange economically even if they lost the drone. Ukrainian forces couldn't afford to let the drones through, the damage to infrastructure was too critical but defending against saturation attacks was bleeding the country's air defense stockpiles at an unsustainable rate.

Ukrainian defenders adapted with characteristic ingenuity. Mobile groups armed with machine guns and MANPADS (man-portable air defense systems) learned to hunt the drones, using their acoustic signature to locate them. The distinctive engine sound that terrorized civilians became a targeting aid for defenders. Teams positioned themselves along likely approach routes, listening for the telltale buzz, then engaging with gunfire when the drones came into visual range.

The Gepard anti-aircraft gun, a Cold War-era German system that Ukraine received in significant numbers, proved surprisingly effective. Its radar could track the Shahed's relatively large radar cross-section, and its twin 35mm cannons could engage at distances and altitudes that made it ideal for drone hunting. What had seemed like obsolete weapons in the age of stealth fighters found new purpose against this low-and-slow threat.

But adaptation had its limits. Russia responded to Ukrainian defensive successes by launching larger waves of drones, spacing them to arrive at different times, approaching from multiple directions simultaneously.

Some drones were fitted with radar reflectors to mimic larger aircraft, drawing air defense fire away from real threats. Others flew low, using terrain masking to avoid radar detection until the last moment.

The Infrastructure War

Through late 2022 and into 2023, Russian Shahed attacks focused relentlessly on Ukraine's energy infrastructure. Power plants and electrical substations became primary targets. Each successful strike didn't just destroy the immediate target, it created cascading failures through Ukraine's interconnected grid. A transformer destroyed in one location meant rolling blackouts across an entire region.

The timing was deliberate. As winter temperatures plunged below freezing, Ukrainian citizens found themselves without heat, without light, without the basic amenities of modern life. Hospitals operated by candlelight and generator power. Families gathered in single rooms, trying to conserve heat. Water pipes froze and burst in unheated buildings. The infrastructure that had taken decades to build was being systematically destroyed in an attempt to make Ukraine ungovernable and Ukrainian life unlivable.

Ukrainian repair crews worked around the clock, often repairing the same facilities multiple times as Russian attacks targeted them again and again. The resilience of Ukraine's energy workers became legendary, men and women working in freezing conditions, sometimes under fire, to restore power to their communities. But heroism alone couldn't overcome the arithmetic of destruction. Repair took time and resources that Ukraine didn't have in infinite supply, while Russia could manufacture and launch Shaheds faster than the damage could be fixed.

International support became crucial. Western nations provided air defense systems, from simple anti-aircraft guns to sophisticated radar-guided missiles. They supplied generators to keep critical infrastructure operating. They transferred technology to help Ukraine detect and track incoming drones more effectively. But the aid, while vital, couldn't completely solve the fundamental problem: Russia was using cheap, expendable weapons to attack targets that were expensive and time-consuming to defend and repair.

The psychological toll mounted alongside the physical destruction. Millions of Ukrainians spent every night listening for air raid sirens, ready to seek shelter at a moment's notice. Sleep deprivation became endemic. Productivity dropped as exhausted workers struggled through their days. Children grew up knowing the sound of explosions, learning to distinguish between different types of impacts, far away, nearby, very close. The normalization of terror became part of daily life.

The Production Race Begins

By December 2022, it was becoming clear that the Shahed threat wasn't temporary. Russia had established a steady supply, and intelligence reports suggested that domestic production facilities were being prepared. Around 2,000 fully assembled drones had been delivered from Iran, but that was just the beginning. Technology transfer agreements meant Russia would soon be manufacturing its own versions, using indigenous resources and supply chains.

Ukrainian intelligence officers studying recovered Shahed wreckage began to understand the full scope of the challenge. The drones' simplicity was actually a strategic advantage. They didn't require sophisticated manufacturing facilities or rare materials. The foam and plywood construction could be accomplished in relatively basic workshops. The engines were adapted from commercial products. The guidance systems, while effective, weren't cutting-edge technology requiring advanced fabrication.

This meant Russia could scale production rapidly once the facilities were established. Estimates suggested that Russia could produce hundreds, then thousands of Shaheds monthly. The leaked documents indicating an expected unit cost of $48,000 for Russian-manufactured versions meant that even with Russia's struggling economy, vast quantities could be afforded. At that price point, even a 50% interception rate still made the attacks cost-effective from Russia's perspective.

Ukraine faced a stark choice: develop countermeasures that could defeat the Shaheds reliably and affordably, or accept that infrastructure attacks would continue indefinitely. Neither option was particularly appealing. Better air defenses required sophisticated technology that Ukraine would need to import or develop domestically, both time-consuming processes. Accepting continued attacks meant accepting the gradual destruction of the country's infrastructure and the constant suffering of its people.

The situation also highlighted a broader shift in the war. This was no longer about battlefield victories or territorial gains in any traditional sense. Russia had effectively abandoned hopes of a quick military victory and had settled into a strategy of attrition, not just attrition of Ukrainian forces, but attrition of the country's infrastructure, economy, and civilian morale. The Shaheds were perfect weapons for this strategy: cheap enough to use in vast numbers, destructive enough to cause meaningful damage, and psychologically impactful enough to terrorize populations.

The International Response

Western intelligence agencies watched the growing Shahed threat with alarm that extended beyond Ukraine. The drones demonstrated how sanctioned nations could acquire military capabilities through technology transfer and domestic production. They showed how relatively primitive technology, employed in sufficient quantities, could create strategic effects despite sophisticated defenses. And they previewed what future conflicts might look like: wars of industrial attrition where cheap, expendable systems were produced faster than they could be destroyed.

The Biden administration announced additional sanctions targeting Iranian entities involved in drone production and transfer. European nations joined in restricting technology exports that could be used in drone manufacturing. But the effectiveness of these measures was limited, the Shaheds didn't rely heavily on cutting-edge Western technology, and the components they did use could be acquired through the complex web of gray market suppliers that sanctions created rather than eliminated.

More practically, Western nations accelerated deliveries of air defense systems to Ukraine. Germany provided more Gepards. The United States sent Avenger systems and committed to Patriot batteries. Norway, Spain, and other nations contributed various anti-aircraft platforms. Each new system helped, but none provided a perfect solution to the fundamental challenge: defending against saturation attacks by cheap expendable drones using expensive, limited-supply interceptors.

Ukrainian defense officials began advocating for a different approach: developing indigenous drone production capabilities that could match or exceed Russia's output. If Russia was going to wage war through mass drone production, Ukraine needed to do the same. The consumer drones and modified commercial systems that had served so well in 2022 were just the beginning. What Ukraine needed was a domestic industry capable of designing, manufacturing, and deploying purpose-built military drones at scale.

The seeds of this industrial transformation were already germinating in workshops and garages across Ukraine and in the international volunteer networks that had sprung up to support the war effort. But transforming ad-hoc production into industrial capacity would take time, resources, and expertise. As 2022 ended and the war entered its second year, the race was on: could Ukraine build its drone industry faster than Russia could scale Shahed production?

The Shape of Things to Come

The arrival of the Shahed marked an inflection point in the drone war. The TB2s that had captured imaginations in the war's opening months were already fading from prominence, their vulnerability to established air defenses limiting their utility. The commercial DJI drones that had revolutionized reconnaissance were still invaluable but were being joined by something new: purpose-built weapons designed specifically for this conflict's unique demands.

The Shahed wasn't sophisticated, but it didn't need to be. It was a guided bomb with wings and an engine, simple enough to mass-produce, reliable enough to reach targets hundreds of kilometers away, and cheap enough to use by the hundreds or thousands. It transformed Russia's strategy from attempting to win the war to attempting to make Ukraine lose by destroying infrastructure, terrorizing civilians, and demonstrating that the suffering would continue indefinitely unless Ukraine capitulated.

But weapons create countermeasures, and strategies create counter-strategies. Ukrainian defenders were learning to hunt Shaheds effectively, developing tactics and techniques that would improve throughout 2023. Ukrainian engineers were studying captured examples, understanding their weaknesses, and planning weapons that could exploit those vulnerabilities. And Ukrainian industry, with international support, was preparing to compete in the mass production race that would define the war's next phase.

As winter descended on Ukraine in late 2022, the Shahed's distinctive buzz became the soundtrack of the war, a nightly reminder that the conflict had entered a new phase where the sky belonged to whoever could build and deploy the most drones, where infrastructure was a battlefield as important as any front line, and where victory would be determined not just by military prowess but by industrial capacity and civilian endurance.

The drone war had been evolving since February, but the Shahed's arrival accelerated that evolution dramatically. What came next would make the developments of 2022 seem almost quaint by comparison. The age of mass drone warfare was beginning, and nobody yet fully understood where it would lead.

Chapter 4

FPV Revolution Begins

The Racer's Edge

In the summer of 2022, while Russian Shahed drones were beginning their terror campaign against Ukrainian cities and DJI Mavics continued their reconnaissance missions along the front lines, a different kind of drone was quietly making its debut on the battlefield. These weren't commercial photography quadcopters or military-grade loitering munitions. They were racing drones, sleek, agile aircraft built for speed competitions where pilots wore video goggles and flew through obstacle courses at breathtaking velocities.

The first documented military use of an FPV (First-Person View) drone in Ukraine occurred in late July 2022, when the volunteer unit "Signum" successfully struck Russian positions. The achievement was captured in a video that would become legendary among drone operators, known simply as "Blue Doors." The footage showed what these adapted racing drones could do: a pilot wearing goggles, seeing through the drone's camera as if sitting in the cockpit, maneuvered the aircraft with precision through a doorway and detonated its payload inside a building occupied by Russian forces.

By August 3, 2022, the 93rd Brigade of Ukraine publicly demonstrated the use of commercial racing drones converted to kamikaze roles. The accompanying text noted that while these drones carried small payloads and were primarily effective against personnel, they represented something fundamentally new: precision strike weapons that cost a few hundred dollars and could be piloted with the kind of fine control that racing drone pilots had been developing for years in competitions.

The racing drone community, both in Ukraine and internationally, recognized immediately what military planners were slower to grasp: these platforms, designed for maneuverability and speed rather than stability and endurance, were perfect weapons for the kind of warfare unfolding in

Ukraine. They were fast, capable of reaching speeds over 100 kilometers per hour in seconds. They were agile, able to dodge, weave, and navigate through complex terrain. And crucially, they were already being built by hobbyists in workshops and garages around the world.

From Hobby to Weapon

FPV racing had emerged as a hobby in the mid-2010s, growing from the broader drone enthusiast community. Unlike the camera drones that most people associated with civilian UAVs, racing drones were built for performance. Lightweight carbon fiber frames, powerful motors, high-capacity batteries, and sophisticated flight controllers combined to create aircraft that could perform aerobatic maneuvers that seemed to defy physics.

The key innovation that made FPV racing possible was the video transmission system. Pilots wore goggles with small screens mounted inches from their eyes, displaying a real-time video feed from a camera mounted on the drone's nose. This created an immersive experience, pilots described it as actually being in the aircraft, feeling every turn and acceleration. The skill required was immense; pilots needed lightning-fast reflexes, three-dimensional spatial awareness, and the ability to process visual information while operating controls with precision.

These same characteristics that made for thrilling racing made for devastating weapons. A skilled FPV pilot could navigate through windows, around obstacles, and into confined spaces with a level of precision that no other weapon system could match. The immersive video feed meant pilots could identify and engage targets with confidence, distinguishing between military vehicles and civilian cars, between occupied positions and abandoned buildings.

The transition from racing to combat required surprisingly few modifications. The basic aircraft design remained the same, carbon fiber frame, brushless motors, lithium polymer batteries, flight controller, video transmitter, and camera. The main addition was a payload: typically a shaped charge explosive or anti-tank grenade mounted to the front of the aircraft. A simple electronic trigger allowed the pilot to detonate the charge at the moment of impact, or in some cases, proximity fuses handled detonation automatically.

The cost differential was staggering. A complete FPV kamikaze drone, including the explosive payload, could be assembled for between $400 and $800, depending on components used. This was orders of magnitude cheaper than any comparable weapons system. A Javelin anti-tank missile cost approximately $175,000. A TB2 drone cost several million dollars. An FPV drone cost less than a decent laptop and could destroy the same targets.

The Volunteer Network Mobilizes

As word spread through the racing drone community about the potential of FPV platforms for military use, volunteers began organizing. Some were Ukrainian nationals already involved in supporting the war effort. Others were international enthusiasts who saw an opportunity to contribute their expertise to a cause they believed in. Racing drone pilots, engineers, and builders formed networks dedicated to producing kamikaze drones for Ukrainian forces.

The beauty of FPV drone production was its accessibility. Unlike sophisticated military systems requiring specialized facilities and rare materials, racing drones could be built anywhere. The components were all commercially available: motors from hobby suppliers, flight controllers from electronics distributors, frames that could be purchased or even

3D-printed, cameras and video transmitters from the same sources that served the racing community.

Workshops sprang up in Ukraine and abroad. In basements and garages, volunteers assembled drones by the dozens. Some operations were remarkably professional, with dedicated assembly lines and quality control procedures. Others were more impromptu—a group of friends gathering on weekends to solder components and test flight controllers. What united them was purpose: every drone they built was one more weapon for Ukrainian soldiers, one more tool that might save lives or destroy enemy equipment.

Khrystyna Pashchenko, a 32-year-old who left her job at an internet firm to volunteer in drone production, described the motivation: "I feel super excited and a little bit proud of myself that I can do something useful." Working unpaid, soldering engines to motherboards, she and thousands like her found meaning in videos from soldiers showing the drones they built in action. The direct connection between their work and battlefield outcomes provided a sense of purpose that no ordinary job could match.

Crowdfunding campaigns, already successful at raising money for commercial reconnaissance drones, expanded to support FPV production. The messaging was compelling: for the price of a nice dinner, donors could fund a drone that might destroy a Russian tank worth millions. The asymmetric economics made the pitch easy. Videos showing FPV drones striking targets, always carefully edited to avoid graphic content while demonstrating effectiveness, went viral attracting more donors and more volunteers.

The Learning Curve

The early FPV kamikaze drones were crude by later standards but effective nonetheless. Pilots learned through trial and error, developing tactics for different scenarios. Attacking armored vehicles required approaching from above, striking at the thin top armor rather than the heavily protected front and sides. Targeting infantry positions meant flying low and fast, using terrain for cover until the final approach. Hitting artillery pieces required patience, waiting for the right moment when crews were exposed.

The video footage from FPV strikes became a crucial tool for training new pilots. Unlike traditional military training that relied on simulations or limited live-fire exercises, FPV pilots could watch dozens or hundreds of real combat missions, seeing exactly what worked and what didn't. Successful tactics spread rapidly through online communities. A pilot in one unit would discover that approaching from a specific angle improved hit rates, and within days, pilots across the front would be using the same technique.

The kill chain for an FPV strike was remarkably short. Reconnaissance drones or ground observers would identify a target and relay coordinates. An FPV pilot would launch, often from positions several kilometers behind the front line. Flight time to target might be just a few minutes. The pilot would acquire the target visually through the FPV camera, maneuver for the optimal attack angle, and strike. From target identification to destruction could take less than ten minutes, far faster than coordinating artillery or air strikes.

But FPV operations weren't without challenges. The video transmission systems that made precision strikes possible also made the drones vulnerable to electronic warfare. Russian forces deployed jammers that could disrupt the radio link between pilot and drone, causing loss of control. Battery life limited range and loiter time, most FPV drones could operate for only ten to fifteen minutes before needing to return for a battery swap. Weather affected operations significantly; wind, rain, and fog all degraded performance or grounded flights entirely.

Pilot training became a bottleneck. Flying an FPV drone required skills that took weeks or months to develop, not the few hours needed to learn DJI Mavic operation. New pilots needed to master basic flight before progressing to more complex maneuvers, then learn combat tactics and target identification. Simulators helped pilots practice on computers before risking expensive hardware but nothing fully replicated the stress and challenges of combat operations.

Russian Response and Adaptation

Russian forces were slow to recognize the FPV threat. In the summer and fall of 2022, their focus remained on defending against TB2 strikes and dealing with reconnaissance drones. FPV attacks were still relatively rare, considered a novelty rather than a serious threat. This complacency gave Ukrainian forces time to refine tactics and scale production before Russia developed effective countermeasures.

By late 2022, however, Russian forces were taking the FPV threat seriously. Electronic warfare units intensified jamming efforts, particularly around high-value targets like command posts and artillery positions. Troops were instructed to watch for small fast-moving drones and engage them with small arms fire when possible. Vehicle crews were told to button up, close hatches and minimize exposed positions when operating near the front lines.

Russia also began developing its own FPV capabilities. Russian pilots, many with backgrounds in the civilian racing community, were recruited and trained. Russian workshops began producing kamikaze drones using similar designs and components to Ukrainian versions. The technology was too simple to keep proprietary; both sides could access the same commercial parts and build similar platforms.

This led to a technological arms race playing out in real-time. Ukrainian engineers developed techniques to make their video transmissions more resistant to jamming. Russian electronic warfare specialists countered with more sophisticated jamming patterns. Ukrainian pilots learned to operate in degraded communications environments, using autonomous flight modes when the video link was disrupted. Russian air defense adapted, developing protocols specifically for engaging small fast-moving targets.

The result was an escalating cycle of innovation and counter-innovation, with both sides constantly adapting to the other's developments. This pace of change, measured in weeks rather than the years typical of traditional military procurement was unprecedented. New tactics and technologies were being developed, tested in combat, refined, and distributed across entire armies faster than doctrine could be written.

The Economic Revolution

The true revolutionary aspect of FPV kamikaze drones wasn't their tactical capability, it was their economics. For the first time in modern warfare, a weapon system existed that was genuinely affordable at scale. Ukraine, a nation with a GDP smaller than many individual U.S. states, could potentially field these weapons in numbers that would overwhelm Russian defenses.

The mathematics were compelling. If Ukraine could produce 1,000 FPV drones per month at $500 each, that represented $500,000 in monthly expenditure. Those 1,000 drones, employed effectively, could destroy targets worth tens or hundreds of millions of dollars: tanks, artillery pieces, armored vehicles, supply trucks, ammunition depots. Even with a 50% loss rate to countermeasures and a 50% miss rate for those that reached targets, 250 successful strikes per month could inflict devastating losses.

Compare this to traditional weapons procurement. A single Javelin anti-tank missile cost $175,000 and destroyed (at best) one vehicle. A thousand Javelins would cost $175 million, 350 times more than a thousand FPV drones and likely wouldn't achieve significantly more destruction given the one-shot nature of both systems. The FPV drones were democratizing precision strike, making it accessible to a nation that couldn't afford to match Russian military spending through conventional means.

This economic calculus wasn't lost on military planners worldwide. If cheap autonomous or semi-autonomous strike weapons could achieve effects previously requiring sophisticated, expensive systems, what did that mean for defense budgets? What did it mean for the balance of military power globally? Small nations, non-state actors, even wealthy individuals could potentially field drone forces capable of threatening conventional military units.

The implications extended beyond immediate military applications. The FPV drone's effectiveness demonstrated that the future of warfare wouldn't necessarily be dominated by whoever could afford the most advanced technology. Instead, it might belong to whoever could produce and field the most platforms, whoever could innovate fastest, whoever could adapt their tactics most quickly to changing conditions.

The Human Element

Behind every successful FPV strike was a human pilot, sitting in a dugout or bunker with goggles strapped to their head, hands on radio controls, guiding a small aircraft through hostile airspace toward a target. These weren't autonomous weapons, they were human-controlled systems requiring constant input and decision-making. The pilot selected the route, identified the target, chose the angle of attack, and decided when to detonate.

This human element was both strength and limitation. Skilled pilots could achieve precision that no autonomous system of the era could match. They could make split-second decisions, abort strikes if civilians appeared, pursue fleeing targets, or adjust tactics based on changing conditions. The immersive FPV experience created a connection between pilot and aircraft that enhanced performance, pilots described being able to "feel" the drone, knowing instinctively how it would respond to control inputs.

But human pilots also meant human limitations. Pilots could fly only so many hours per day before exhaustion degraded performance. They needed training, rest, rotation. They experienced stress, fear, and the psychological burden of killing, even when done remotely through a video screen. Some pilots struggled with the intimate nature of FPV combat, watching through the drone's camera as it approached a target, seeing enemy soldiers in detail seconds before impact.

The FPV community developed its own culture and psychology. Pilots shared videos, comparing strikes and discussing tactics. They developed nicknames and reputations based on their skill and effectiveness. Some became almost competitive, tracking their confirmed kills as if keeping score. Others were more somber, recognizing that each successful strike meant human death, however necessary in the context of war.

Looking Forward

By the end of 2022, FPV kamikaze drones remained a relatively niche capability, employed by specialized units and volunteer groups rather than integrated throughout Ukrainian forces. Production numbered in the hundreds per month rather than thousands. Many military officers remained skeptical, viewing them as supplements to conventional weapons rather than potential replacements.

But the foundation was being laid for something much larger. The volunteer networks were growing more sophisticated, production processes were being refined, and more pilots were being trained. Word was spreading through the military about units achieving remarkable success with FPV drones. Videos of strikes were going viral online, attracting attention from senior commanders and political leaders.

Ukrainian President Volodymyr Zelensky had taken notice of the FPV phenomenon. His government was beginning to consider how these grassroots efforts could be scaled, systematized, and integrated into the broader war effort. The question wasn't whether FPV drones would play a larger role, it was how large that role could become and how quickly Ukraine could scale production to meet the enormous potential demand.

In workshops across Ukraine and in diaspora communities around the world, volunteers continued building drones by hand, one at a time. Each drone represented hope, hope that technology and innovation could help offset Russia's advantages in manpower and traditional military equipment. They couldn't know it yet, but they were building not just weapons but the foundation of a new way of war, one where the sky would belong to swarms of small, cheap, expendable aircraft guided by human skill and determination.

The FPV revolution had begun quietly, almost stealthily, while the world's attention focused on TB2s and Shaheds and conventional battles. But in the laboratories of combat, where tactics are tested under the most demanding conditions imaginable, FPV drones were proving themselves. The lessons learned in 2022 would shape 2023's dramatic expansion, setting the stage for the true age of mass drone warfare. The future was being built in garages and basements, soldered together by volunteers, and proven in the skies over Ukraine's battlefields.

PART II: INDUSTRIAL WARFARE (2023)

Chapter 5

Scaling Up Production

The winter of 2022-2023 brought a sobering realization to Ukraine's military leadership: the drone war wasn't going to end anytime soon, and the country's dependence on imported commercial drones represented a strategic vulnerability that could prove fatal. As Russia began mass-producing Shahed kamikaze drones with Iranian assistance and both sides burned through hundreds of reconnaissance drones monthly, Ukraine faced an existential choice, build a domestic drone industry virtually from scratch, or watch as supply shortages handed Russia an insurmountable advantage.

What followed was one of the most remarkable industrial mobilizations of the 21st century, a transformation that saw Ukraine evolve from a drone importer to a drone manufacturing powerhouse in less than eighteen months. This wasn't the carefully planned, centrally coordinated industrial mobilization of World War II. Instead, it was chaotic, distributed, innovative, and ultimately effective, a uniquely Ukrainian response to an unprecedented challenge.

The Garage Workshop Beginning

In early 2023, Ukraine's drone production was still primarily the domain of volunteers, hobbyists, and small workshops scattered across the country. Maksym, a software engineer from Lviv who had spent the first year of the war sourcing DJI drones from Poland and Slovakia, made a decision that would be replicated hundreds of times across Ukraine. He stopped trying to import drones and started building them.

His workshop occupied a corner of his apartment building's basement, three tables, basic soldering equipment, 3D printers running continuously, and bins filled with motors, flight controllers, and carbon fiber frames ordered from Chinese suppliers. Within two months, he and four volunteers were producing fifteen FPV drones weekly. The drones were crude by commercial standards, with exposed wiring and hand-soldered connections, but they flew, they carried explosives, and most importantly, they worked.

"We had no idea what we were doing at first," Maksym recalled in an interview conducted in summer 2023. "I'd built maybe two racing drones before the war as a hobby. Now we were trying to make weapons. But the soldiers didn't care if the soldering looked pretty. They cared if it would fly into a Russian tank and explode."

Maksym's operation was microscopic compared to what was coming, but it represented the seed of Ukraine's distributed manufacturing model. Across the country, similar workshops sprouted in garages, apartments, evacuated shops, and rural barns. A mechanic in Dnipro started producing drone frames from salvaged aluminum. A Kyiv electronics teacher began training teenagers to assemble flight controllers. In Odesa, a former toy factory pivoted to producing quadcopter parts.

These weren't coordinated efforts, at least not initially. They were organic responses to immediate needs, driven by soldiers at the front sending desperate messages to friends and contacts: "We need more eyes in the sky. We need something that can hit their positions. Can you help?"

The Volunteer Networks Scale Up

By March 2023, the scattered workshop efforts began to coalesce into something resembling a supply network. Volunteer organizations that had spent 2022 fundraising to purchase commercial drones now redirected their efforts toward domestic production. The mathematics were compelling: a DJI Mavic 3, assuming one could even find them for sale after China's export restrictions tightened, cost $2,000-3,000. A domestically assembled FPV kamikaze drone cost $300-500 in parts.

The volunteer networks became the connective tissue of Ukraine's emerging drone industry. Organizations like the "Angry Birds" collective, which had raised millions through crowdfunding campaigns, began contracting with multiple small manufacturers, providing startup capital and guaranteed purchase orders. This solved the crucial chicken-and-egg problem: workshops needed money to buy equipment and components, but they couldn't earn money until they were producing at scale.

Olena, who ran a volunteer network coordinating between seven different workshops in western Ukraine, described the system that emerged: "We'd get requests from specific units, 'We need 50 reconnaissance drones and 200 FPV kamikaze drones this month.' Then we'd divide the orders among our workshops based on their capabilities and capacity. One workshop specialized in long-range reconnaissance platforms. Another could churn out kamikaze drones faster than anyone. A third was experimenting with new designs. It wasn't efficient by peacetime standards, but it was resilient. If one workshop got bombed or lost power, the others kept producing."

This distributed model had unintended advantages. Unlike centralized factories, which made tempting targets for Russian missiles, the garage workshops were nearly impossible to target effectively. They had minimal physical footprints, could relocate quickly, and were scattered across the country. Destroying Ukraine's drone production would require thousands of precision strikes on small, mobile targets, an impossible task even for Russia's most capable systems.

From Hobby Shop to Assembly Line

As spring turned to summer in 2023, the most successful workshops began their transformation from hobby operations to something approaching real manufacturing facilities. This evolution was driven partly by increasing demand, combat units that had received a dozen drones in January were requesting hundreds by June and partly by the workshops' own learning curves.

The transformation of the "Wild Hornets" workshop in Cherkasy illustrates the progression. Founded by three friends with backgrounds in engineering and electronics, Wild Hornets started in a two-car garage in February 2023, producing about twenty drones monthly. By August, they had moved to a small industrial building, employed thirty-two people, and were producing 400 drones monthly.

The key was systematization without losing flexibility. Andriy, one of the founders, explained their evolution: "In the beginning, every drone was basically a custom build. Someone would assemble it from start to finish, testing each component as they went. It worked, but it was slow. We started breaking the process into stations, one team would prepare the frames, another would install the motors and ESCs, another would program the flight controllers, another would do final assembly and testing. Production tripled within a month."

But Wild Hornets, like most Ukrainian workshops, resisted the temptation to standardize everything. Combat feedback was arriving constantly, soldiers reporting that a certain motor failed too often in cold weather, or that the camera angle needed adjustment for urban environments, or that they needed a variant that could carry a heavier warhead. The ability to iterate rapidly, to implement changes within days rather than months, became a strategic advantage.

"Russian drone production was more organized, more standardized," Andriy noted. "But that made them slower to adapt. When their Lancet drones started getting jammed more effectively, it took them months to deploy countermeasures. When our drones got jammed, we'd have a new solution being tested within a week."

The Combat Unit Manufacturers

Perhaps the most unusual aspect of Ukraine's drone production surge was the emergence of combat units manufacturing their own drones. By mid-2023, dozens of frontline units had established their own small-scale production capabilities, usually producing 20-50 drones monthly for their specific operational needs.

These unit-level workshops solved a critical problem: the lag time between identifying a tactical need and receiving a solution. When a reconnaissance unit discovered that Russian forces were using a particular frequency for communications, they needed drones equipped to intercept those signals immediately, not after a request worked its way through procurement channels. When an assault unit needed drones optimized for urban warfare in Bakhmut's ruins, they built them.

Lieutenant Dmytro, whose mechanized battalion established a small workshop in a liberated village near the Zaporizhzhia front, described the advantages: "We'd test something in combat in the morning, and by evening, the workshop guys were modifying the design. Our drones were ugly, inconsistent, and probably violated every manufacturing standard that existed. But they were exactly what we needed for our specific situation."

The unit-level production also created an unofficial but highly effective research and development network. Innovations developed by one unit would spread through informal channels, soldiers on leave visiting other units, mechanics comparing notes, videos of successful tactics spreading through military Telegram channels. A clever modification developed in Kherson would appear in workshops in Donetsk within weeks.

This grassroots innovation network operated entirely outside official channels, driven by the imperative of combat effectiveness rather than bureaucratic procedure. It was messy, redundant, and sometimes led to dead ends. But it was also remarkably adaptive, generating hundreds of small improvements that collectively transformed Ukrainian drone capabilities.

The Absence of Central Planning

The Ukrainian government's role in this industrial surge was notable primarily for what it didn't do. There was no Five Year Plan for drone production, no centralized allocation of resources, no ministry coordinating all efforts. Instead, the government's contribution was largely to get out of the way.

In peacetime, Ukraine's bureaucracy would have strangled the garage workshops with regulations. Manufacturing weapons required licenses, safety certifications, quality control inspections, and endless paperwork. Environmental regulations would have complicated chemical handling for explosives. Labor laws would have complicated the volunteer workforce. Import/export controls would have delayed component shipments.

In wartime, most of these rules were simply ignored or quietly set aside. The government's unspoken policy was that if drones were reaching the front and soldiers were using them effectively, no one was going to shut down production over permit violations. This regulatory gray zone became a crucial enabler of rapid scaling.

"We were technically operating illegally for the first six months," admitted Kateryna, who ran a Kyiv workshop that grew to employ forty people by late 2023. "We were manufacturing weapons without a license, importing components through customs channels that weren't really designed for military equipment, and paying our workers in cash to avoid paperwork delays. If this had been peacetime, we would have been shut down immediately. But everyone understood the situation. The customs officers, the local police, even the tax authorities, they all knew what we were doing and why. Nobody asked difficult questions."

This wasn't complete chaos. The government did establish some basic frameworks, systems for workshops to register as military suppliers, simplified customs procedures for drone components, tax incentives for domestic production. But these were enabling structures rather than controlling mechanisms. The government created conditions for production to flourish rather than attempting to direct it centrally.

The approach had drawbacks. There was significant redundancy, with multiple workshops solving the same problems independently. Quality control was inconsistent, some workshops produced excellent drones while others' products failed embarrassingly often. There were cases of fraud, with workshops taking funding and producing little. Resources weren't allocated optimally, with some promising manufacturers struggling for capital while less capable operations received generous funding from enthusiastic but naive donors.

But the advantages outweighed the inefficiencies. The distributed model was resilient, adaptive, and scalable in ways that centralized production couldn't match. When the system worked and increasingly it did, it produced thousands of drones monthly from dozens of sources, each iterating and improving based on combat feedback.

The Chinese Component Challenge

Ukraine's drone surge faced one critical vulnerability: almost every workshop depended on components imported from China. Motors, electronic speed controllers, flight controllers, cameras and batteries the vast majority came from Chinese manufacturers, often the same suppliers that served the global hobbyist and commercial drone markets.

This created constant anxiety about supply chain security. What if China imposed strict export controls? What if suppliers refused to sell to Ukrainian buyers? What if shipping routes were disrupted?

In practice, these fears were only partially realized. China did impose some restrictions, particularly on complete drone systems like DJI products. But the component-level supply chain proved remarkably resilient. Chinese manufacturers, many of them private companies motivated primarily by profit, continued selling to Ukrainian buyers through various channels. Components that couldn't be imported directly came through Poland, Turkey, and other third countries. Some workshops established shell companies to obscure the end destination of their orders.

The workshops also began strategic stockpiling. By mid-2023, successful manufacturers were maintaining 2-3 months of component inventory, cushioning against supply disruptions. Some started qualifying multiple suppliers for critical components, accepting slightly higher costs for supply security.

There were also modest efforts at component localization. A Kharkiv factory began producing batteries that met 60-70% of domestic demand by late 2023. Several workshops developed relationships with Ukrainian electronics manufacturers to produce flight controller boards, though Chinese imports remained cheaper and more capable. Motors proved particularly difficult to source domestically, specialized brushless motors for drones required manufacturing capabilities that Ukraine largely lacked.

The dependence on Chinese components remained a strategic vulnerability, but it was a manageable one. The supply chain proved robust enough to support the production surge, even as workshops and government officials worried constantly about its fragility.

The Numbers Tell the Story

By December 2023, the transformation was undeniable in the production numbers. Ukraine's domestic drone manufacturing had grown from essentially zero in January to an estimated 50,000-80,000 drones monthly by year's end. The exact figures were treated as military secrets, but industry estimates and occasional official statements painted a picture of explosive growth.

TAF Drones, which would later become Ukraine's largest manufacturer, was already producing several thousand units monthly. Dozens of mid-sized workshops were each churning out hundreds of drones. Hundreds of smaller operations contributed dozens each. Even accounting for the cottage industry and unit-level production, the total was staggering.

More importantly, production was still accelerating. Workshops that had been producing for six months were doubling their output as they refined their processes. New workshops were opening monthly. International funding was flowing in at increasing volumes as donor nations recognized the strategic importance of Ukraine's drone capabilities.

The drones themselves were also evolving. The crude, hand-assembled FPV kamikaze drones of early 2023 had given way to more sophisticated designs with better range, more reliable electronics, and improved warhead integration. Specialized variants were emerging, reconnaissance drones with thermal cameras, long-range strike drones, urban warfare optimized platforms and electronic warfare drones.

Ukraine had built a drone industry under fire, transforming from complete dependence on imports to substantial self-sufficiency in twelve months. It was inefficient by peacetime standards, unorthodox by military procurement norms, and probably unsustainable in the long term. But it worked.

As 2024 approached, the question was no longer whether Ukraine could produce enough drones. It was whether Russia could keep pace and whether either country's economy could sustain the consumption rates that mass drone warfare demanded. The race was on, and Ukraine had proven it could run.

Chapter 6

The Battle for the Skies

The explosion of drone production in 2023 triggered an equally dramatic escalation in the invisible war fought across the electromagnetic spectrum. For every drone launched, there was an electronic warfare system trying to prevent it from reaching its target. For every innovation in guidance and control, there was a countermeasure designed to defeat it. The battle for the skies wasn't just physical, it was electronic, and the stakes were nothing less than determining whether drones could function as effective weapons or expensive flying debris.

By mid-2023, Ukrainian drone operators faced a grim reality: more drones were being lost to electronic warfare than to traditional air defenses or small arms fire. Russian jamming systems had become so effective that in some sectors of the front, Ukrainian drone operations were reduced to less than 30% success rates. The crisis forced a fundamental rethinking of how drones operated, how they were controlled, and what technologies they could rely upon in a saturated electronic warfare environment.

This was a war without front pages or dramatic footage, fought by engineers and electronic warfare specialists rather than infantrymen. But it would prove as decisive as any artillery barrage or armored assault.

The Jamming Escalation

The early months of the war had seen relatively primitive electronic warfare efforts against drones. Russian forces employed commercial signal jammers, often the same equipment sold to prisons and security services to block cell phone signals. These systems could disrupt the 2.4 GHz and 5.8 GHz frequencies that commercial drones used for control and video transmission, but their range was limited and their effectiveness sporadic.

By early 2023, the situation had changed dramatically. Russia deployed increasingly sophisticated military-grade electronic warfare systems specifically designed to counter drone operations. The Pole-21 jamming system, originally designed to disrupt Ukrainian communications, was redeployed to target drone frequencies. The Borisoglebsk-2 electronic warfare complex, one of Russia's most capable systems, began dedicating significant resources to drone interdiction.

These weren't simple signal blockers. They were intelligent systems capable of identifying drone transmissions, analyzing their characteristics, and targeting them with focused jamming that could overwhelm the drone's receiver with noise. The effect on Ukrainian drone operations was devastating.

Yuri, an FPV drone operator with the 93rd Mechanized Brigade, described his experience in March 2023: "We'd launch a drone, and for the first kilometer everything would be fine. Then suddenly, nothing. Total loss of control. The video feed would freeze or turn to static. The drone would either crash or activate its failsafe and return to the launch point, which meant it was useless for attack missions. We were losing seven or eight drones out of every ten we launched, and most of those losses were to jamming, not enemy fire."

The crisis was compounded by the reliance on GPS for navigation. Most Ukrainian drones, especially the domestically produced FPV platforms, used GPS coordinates for navigation and autonomous flight modes. Russian GPS jamming and spoofing systems proved remarkably effective at denying this capability. Drones would suddenly believe they were hundreds of kilometers from their actual location, or would lose position data entirely and become uncontrollable.

The mathematics of the jamming war favored the defender. A jammer could protect a large area, sometimes several square kilometers while each drone represented a single point target trying to penetrate that protection. Russia's electronic warfare doctrine, developed over decades and refined in

Syria, gave them sophisticated playbooks for defeating aerial systems. They were prepared for this fight in ways that Ukraine initially was not.

The Cat-and-Mouse Game Begins

The Ukrainian response to the jamming crisis came from multiple directions simultaneously. Military signals intelligence units, commercial drone manufacturers, volunteer engineers, and frontline operators all contributed pieces of the solution. What emerged was not a single breakthrough but rather a continuous cycle of adaptation, a technological arms race compressed into weeks and months rather than years.

The first wave of counter-jamming measures focused on frequency hopping and increased transmission power. Ukrainian workshops began installing more powerful transmitters in their drones, boosting signal strength to overcome jamming. They implemented frequency hopping spread spectrum (FHSS) systems that constantly switched transmission frequencies, making it harder for jammers to lock onto and disrupt the signal.

Vadym, an electronics engineer who had pivoted from designing commercial WiFi equipment to modifying military drones, explained the technical challenge: "The Russians had powerful jammers that could blanket entire frequency bands with noise. Our solution was to use more of the available spectrum, switching frequencies hundreds of times per second. It was like trying to have a conversation in a noisy room by constantly changing languages. The noise is still there, but if you're switching faster than your opponent can adapt, you can get your message through."

These modifications helped but weren't decisive. Russian electronic warfare units adapted their jamming patterns to target the specific frequency hopping sequences Ukrainian drones used. The cycle would repeat, Ukraine would modify their hopping algorithms, Russia would adjust their jamming, Ukraine would respond with new modifications.

Directional antennas became another crucial tool. Standard omnidirectional antennas radiated signal in all directions, making them easy targets for jammers. Ukrainian operators began using high-gain directional antennas that focused signal strength toward the drone, making the link more resilient to interference. This required operators to track the drone's position and manually aim the antenna, adding complexity but improving success rates significantly.

By summer 2023, successful Ukrainian drone operators looked like a cross between pilots and radio operators, surrounded by multiple antennas, signal strength meters, and spectrum analyzers, constantly adjusting their equipment to maintain the fragile link with their aircraft.

GPS Spoofing and the Navigation Crisis

While jamming disrupted control links, GPS spoofing represented an even more insidious threat. Rather than simply blocking GPS signals, Russian electronic warfare systems transmitted false GPS signals that convinced drones they were in completely different locations. The effects ranged from confusing to catastrophic.

Captain Oleksandr, who commanded a drone reconnaissance unit in the Zaporizhzhia sector, described the problem: "We'd program a drone to fly to coordinates that we knew were Russian positions, we could see them with our own eyes three kilometers away. But the drone would veer off course, sometimes flying in completely wrong directions. We had drones that thought they were in Belarus when they were actually over Donetsk. The GPS spoofing was so sophisticated that the drone's navigation system would accept the false data as legitimate."

The spoofing went beyond simple position errors. Some Russian systems generated moving false coordinates that would guide Ukrainian drones away from sensitive targets and into kill zones where anti-aircraft systems waited. Others created virtual "GPS walls" where drones would suddenly believe they had reached the boundaries of their operational area and automatically return to base.

Ukrainian counter-measures evolved along multiple paths. The most sophisticated solution involved inertial navigation systems (INS) that didn't rely on GPS at all. These systems used accelerometers and gyroscopes to track the drone's movement relative to its starting point, similar to the technology in intercontinental ballistic missiles. Combined with optical flow sensors that tracked ground features, INS-equipped drones could navigate even in complete GPS denial.

But INS systems were expensive, complex, and required precise calibration. They couldn't be retrofitted to the thousands of cheap FPV drones that formed the backbone of Ukrainian drone forces. For these systems, simpler solutions emerged.

Visual navigation became increasingly important. Experienced operators learned to fly manually using only the drone's camera feed, navigating by landmarks and terrain features. This required significant skill, flying an FPV drone at high speed through complex terrain while evading jamming was compared to playing a high-stakes video game where failure meant losing a $500 piece of equipment and potentially compromising a military operation.

Some workshops implemented dual GPS receivers that could compare signals from multiple satellites and detect spoofing attempts. Others integrated compass-based navigation as a backup, though this was vulnerable to magnetic interference. The most creative solution involved using visual landmarks, operators would mark distinctive terrain features visible from the air, and drones would navigate using these references combined with dead reckoning.

The Range Problem

Electronic warfare fundamentally altered the tactical calculus of drone operations. In a clean electromagnetic environment, an FPV drone with a well-designed control system could operate at ranges of 10-15 kilometers, giving operators standoff distance from frontline dangers. Under intense jamming, that range collapsed to 2-3 kilometers or less, forcing operators to work dangerously close to the front lines.

This created a cascading set of problems. Operators closer to the front were more vulnerable to counter-battery fire, infantry raids, and Russia's own drone attacks. The compressed operating range made it easier for Russian forces to locate launch sites through direction-finding equipment. And missions that had been routine, striking logistics hubs, command posts, or artillery positions well behind the front lines became extremely difficult or impossible.

Ukrainian manufacturers responded by developing drones optimized for jamming environments. These platforms sacrificed some performance characteristics for improved electronic resilience. They used lower frequency control links that were harder to jam but required larger antennas. They incorporated redundant control systems that could switch between different communications methods if the primary link failed. They added autonomous features that allowed drones to complete missions even after losing contact with operators.

The most significant development was the emergence of truly autonomous operation. Rather than maintaining constant control links, operators would program complete mission profiles into the drone before launch, flight path, target identification criteria, attack parameters. The drone would execute the mission independently, transmitting data only intermittently or not at all until returning to base.

This autonomy raised effectiveness but introduced new challenges. Autonomous drones were harder to recall if circumstances changed or if friendly forces entered the target area. They required sophisticated target recognition systems to ensure they struck the right objectives. And they raised ethical and legal questions about weapons systems operating without continuous human control.

By late 2023, the most advanced Ukrainian drones operated in a hybrid mode, autonomous navigation with periodic check-ins for operator confirmation before final attack. This balanced effectiveness in jamming environments with maintaining human control over lethal decisions.

The Frequency Hunters

As the electronic warfare battle intensified, both sides deployed specialized units whose sole purpose was to hunt enemy drones and operators through the electromagnetic spectrum. These frequency hunting teams combined electronic intelligence equipment, direction-finding systems, and quick-reaction strike capabilities to target drone operations.

Russian frequency hunters became particularly effective. They would detect Ukrainian drone transmissions, use triangulation to locate the operator's position within minutes, and call in artillery or Lancet loitering munitions on the site. Ukrainian drone operators learned through brutal experience that extended operations from a single location were suicidal.

Lieutenant Dmytro, who survived four artillery strikes on his operating positions, described the deadly routine: "We'd set up, launch a few drones, and then pack up and move immediately. Never more than 15 minutes in one location. The Russians had the system down to a science, detect, locate, strike. We lost three operators in my unit before we learned to keep moving constantly."

This forced Ukrainian drone units into a nomadic existence, operating from the backs of trucks or from rapidly prepared positions that could be abandoned within minutes. Some units developed elaborate deception plans, transmitting from decoy locations to draw strikes away from actual operating positions. Others used remote transmission systems, operating their drones from positions kilometers away from the actual transmitter.

The frequency hunt worked both ways. Ukrainian signals intelligence units became adept at locating Russian electronic warfare systems and drone operators. The signature of a powerful jammer was difficult to hide, and Russian EW units became priority targets for Ukrainian artillery and long-range drones. Several high-value Russian electronic warfare systems were destroyed after being located through their own transmissions.

This created a complex calculation for both sides. Activating powerful jamming systems made drone operations nearly impossible but also revealed the jammer's location and invited attack. Sophisticated Russian EW units learned to operate in pulse mode, activating their systems for brief periods, relocating, and activating again from new positions. This reduced their effectiveness but improved survivability.

The Antenna Arms Race

As signals warfare intensified, the antennas themselves became sophisticated pieces of military technology. The standard commercial antennas that came with hobby drones were hopelessly inadequate for military operations under jamming. Ukrainian workshops and specialized manufacturers developed increasingly exotic antenna designs optimized for different tactical situations.

Helical antennas provided circular polarization that improved signal penetration through jamming. Patch antenna arrays offered high gain in compact packages that could be mounted on vehicles. Yagi antennas delivered maximum range for fixed-position operations. Cross-polarized systems could maintain links even when drones maneuvered dramatically.

The most sophisticated Ukrainian drone units deployed antenna systems that cost more than the drones themselves, custom-built arrays with automatic tracking, signal amplification, and filtering systems to extract drone signals from overwhelming electromagnetic noise. These systems required dedicated operators and technical specialists, transforming drone operations from something a single soldier could manage to a team effort requiring multiple personnel and substantial equipment.

Roman, who built custom antenna systems for Ukrainian special operations units, explained the evolution: "In 2022, guys were using antennas that came in the box with Chinese racing drones. By 2023, we were building custom systems with up to 25 decibels more gain, multiple polarizations, active filtering, and automatic tracking. The difference in performance was night and day. A drone that would lose connection at 2 kilometers with a stock antenna could operate at 8-10 kilometers with our custom systems, even under moderate jamming."

But specialized antennas came with trade-offs. They were bulkier, heavier, and more fragile than standard equipment. They required training to operate effectively. And they were expensive, a complete antenna system for a single drone operator might cost $3,000-5,000, many times the cost of the drones themselves.

This created a two-tier system. Elite units with substantial resources operated with sophisticated antenna systems and modified drones that maintained effectiveness even under intense electronic warfare. Regular units made do with simpler equipment, accepting lower success rates and shorter range in exchange for lower costs and simpler operations.

The Survivability Imperative

By the end of 2023, drone survivability in contested electromagnetic environments had become the paramount design criterion. Range, speed and payload capacity all these factors mattered less than whether a drone could complete its mission when Russian electronic warfare systems were active.

Ukrainian manufacturers responded with radical design changes. Some platforms abandoned traditional radio control entirely in favor of optical fiber guidance, a return to technologies from the 1970s that proved remarkably effective against 21st-century jamming. Others implemented multiple redundant control systems using different frequencies and protocols, ensuring that if one system was jammed, others might still function.

Machine vision systems became increasingly important. Rather than relying on GPS coordinates or operator guidance, advanced drones used cameras and computer vision to identify targets autonomously. A drone might be given a rough mission profile, "fly north 5 kilometers and attack the largest vehicle you find" and execute that mission entirely on its own, immune to jamming because it never needed external guidance.

These autonomous systems were expensive and complex, but they proved their worth in the most heavily jammed sectors of the front. Where traditional drones had success rates below 30%, autonomous platforms maintained 60-70% mission success even under intense electronic warfare.

The survivability focus transformed Ukrainian drone doctrine. Missions were planned assuming degraded or absent communications. Operators were trained to program complete autonomous sequences rather than flying manually. Drones were designed with the assumption that they would spend most of their flight time beyond radio contact, returning to controlled flight only at the beginning and end of missions.

The Lessons Learned

The electronic warfare battle of 2023 taught both sides harsh lessons about modern drone warfare. The comfortable assumption that cheap commercial technology could be easily militarized proved false or at least incomplete. Yes, drones could be built cheaply, but making them survivable against sophisticated electronic warfare required substantial investment in counter-measures, training, and supporting equipment.

For Ukraine, the electronic warfare challenge drove innovation in ways that might not have occurred otherwise. The necessity of operating under jamming forced the development of more sophisticated autonomous systems, better navigation technologies, and more resilient communications. These capabilities would prove crucial as the war continued and Russian electronic warfare became even more capable.

For Russia, the battle demonstrated that electronic warfare, while extremely effective, wasn't a complete solution. Even with significant electronic warfare advantages, Russian forces couldn't stop Ukrainian drones entirely. The Ukrainians proved remarkably adaptive, finding workarounds and counter-measures faster than Russian doctrine anticipated. The war became an exhausting cycle of measure and countermeasure with no end in sight.

The implications extended far beyond the Russia-Ukraine conflict. Military planners worldwide studied the electronic warfare lessons carefully. The comfortable Western assumption that advanced technology would dominate any battlefield was called into question. If Russian electronic warfare could suppress sophisticated Ukrainian drones, what would it do to Western systems that had never been tested under such conditions?

As 2024 approached, the electromagnetic battlefield remained hotly contested, neither side achieving decisive dominance. The drones still flew, missions still succeeded, but only through constant innovation and adaptation. The battle for the skies continued, invisible to cameras but absolutely critical to the war's outcome. In this war, the side that controlled the electromagnetic spectrum would ultimately control the physical battlefield and both sides knew it.

Chapter 7

Russia's Shahed Factory

While Ukraine was building its drone industry from the ground up through distributed workshops and volunteer networks, Russia pursued a very different path to drone dominance. Their solution came not from domestic innovation but from an unlikely alliance with Iran, and eventually, the transfer of entire production capabilities to Russian soil. The story of the Shahed-136 drone, known to Russians as the Geran-2 represents one of the most significant arms transfers of the 21st century and transformed Russia's ability to wage long-range strategic warfare.

The journey from imported Iranian weapons to mass domestic production revealed both Russia's industrial capabilities and its vulnerabilities. It demonstrated how quickly drone technology could proliferate, how difficult it was to prevent even with sanctions and export controls, and how a relatively simple design could have devastating strategic effects when produced at scale.

The Iranian Lifeline

By summer 2022, Russia faced a growing crisis in its long-range strike capabilities. Precision-guided missiles, Kalibr cruise missiles, Iskanders, Kh-101 air-launched weapons were being expended at unsustainable rates. These weapons cost millions of dollars each, and Russian production capacity, constrained by Western sanctions on critical components, couldn't replace them fast enough. Russia needed a cheaper way to strike targets deep inside Ukraine.

Iran offered a solution. The Islamic Revolutionary Guard Corps had spent years developing unmanned aerial systems for asymmetric warfare, and their Shahed-136 loitering munition represented a mature, combat-proven design. Crucially it was cheap, estimates suggested production costs of $20,000-50,000 per unit compared to millions for cruise missiles. And it could be built with components that weren't subject to the same export restrictions as advanced missiles.

The first evidence of Iranian involvement came in mid-August 2022, when Ukrainian forces shot down an unfamiliar drone near Kupiansk. Analysis revealed a delta-wing design with a pusher propeller, constructed primarily from foam and plywood. The engine was a commercial Chinese gasoline motor of the type used in motorcycles and small generators. The guidance system used commercial GPS and a simple inertial navigation system. The warhead contained approximately 40 kilograms of high explosives.

It was crude by Western standards more like a large model aircraft than a precision weapon. But it flew up to 2,000 kilometers, it could loiter for hours searching for targets, and most importantly, it could be produced in large numbers quickly and cheaply. The Shahed-136 wasn't elegant, but it was effective.

Intelligence estimates suggest Iran initially delivered approximately 2,000 Shahed-136 drones to Russia between August and December 2022. These weapons were used primarily for terror attacks on Ukrainian cities and infrastructure, power stations, water facilities, and civilian heating plants. The goal wasn't military advantage but psychological warfare and the systematic degradation of Ukrainian civilian morale through constant attacks.

The initial Iranian deliveries, however, were just the beginning. Russia's real objective was acquiring the capability to produce these weapons domestically, in quantities that would dwarf the initial shipments.

Technology Transfer

The transfer of production technology from Iran to Russia represented a complex undertaking that violated multiple international agreements and required elaborate sanctions evasion. Iranian engineers traveled to Russia under various covers, tourism, business ventures and academic exchanges. Russian technical delegations visited Iranian production facilities, ostensibly for agricultural or commercial purposes. Documents, blueprints, and specifications moved through encrypted channels and human couriers.

Western intelligence agencies tracked portions of this technology transfer but proved unable to prevent it. The knowledge being transferred wasn't cutting-edge technology requiring exotic materials or ultra-precise manufacturing. The Shahed-136 was deliberately designed for producibility, using widely available components and relatively simple construction techniques. This made it nearly impossible to interdict through export controls.

By early 2023, Russia had established production facilities in multiple locations. The primary manufacturing site was reportedly near Yelabuga in Tatarstan, though intelligence reports suggested additional facilities in other regions. The distributed production model reminiscent of Ukrainian drone manufacturing, though on a more industrial scale made the facilities difficult to target and ensured that disruption at one site wouldn't cripple the entire program.

The Russian production version, designated Geran-2, incorporated modifications to the Iranian design. Russian engineers made improvements to the guidance system, enhanced the engine for better reliability in cold weather, and modified the warhead to use Russian explosives. Some versions incorporated Russian electronic warfare countermeasures. The basic design remained Iranian, but Russian manufacturing techniques and available components shaped the production version.

Establishing domestic production wasn't without challenges. The Shahed design relied heavily on Western electronics and Chinese components. Microcontrollers, GPS modules, communication systems, and various electronic components had to be sourced through elaborate supply chains designed to evade sanctions. Russian procurement networks, operating through front companies in Turkey, Dubai, Hong Kong, and other commercial hubs, worked to secure the thousands of components needed for mass production.

The Smuggling Network

The component supply chain that fed Russia's Shahed production represented a masterclass in sanctions evasion. Western electronics, chips from American manufacturers, sensors from European companies, communication modules from Japan ended up in Russian drones despite extensive export restrictions.

The routes were circuitous and constantly shifting. A microcontroller might be manufactured in Taiwan, shipped to a distributor in Singapore, sold to a trading company in Dubai, resold to a Turkish electronics wholesaler, transferred to a shell company in Kazakhstan, and finally trucked into Russia labeled as civilian goods. Each step provided plausible deniability and obscured the ultimate destination.

Chinese components presented less challenge but still required careful handling. While China maintained official neutrality and claimed to restrict military exports to belligerents, Chinese commercial entities proved remarkably willing to supply components that clearly had military applications. The small gasoline engines that powered the Shaheds, for instance, came from Chinese manufacturers who asked few questions about why a Russian buyer needed thousands of 50cc two-stroke motors.

Western intelligence agencies and investigative journalists documented numerous cases of sanctioned components appearing in downed Russian drones. One particularly detailed analysis of a Shahed recovered intact in Ukraine identified parts from over thirty different manufacturers across a dozen countries. Many were obsolete or commodity components not cutting-edge technology, but the mundane building blocks of modern electronics that were produced in vast quantities and sold through global supply chains that were nearly impossible to police effectively.

The Iranian connection provided additional supply chain benefits. Iran, long accustomed to operating under sanctions, had developed sophisticated procurement networks that could be leveraged for the Russian program. Iranian front companies with established relationships with suppliers worldwide helped Russian buyers access components that direct Russian purchases would have struggled to obtain.

The economics of component procurement shaped production rates. When supply chains functioned smoothly, production could surge. When Western sanctions enforcement tightened or key suppliers became cautious, production would dip. Throughout 2023, Russian Shahed production ebbed and flowed with the health of the smuggling network, but the overall trend was steadily upward.

Scaling Production

By late 2023, Russian Shahed production had reached industrial scale. While exact production numbers remained classified, analysis of attack patterns and recovered drones suggested Russia was producing several hundred Shaheds monthly, far exceeding the initial Iranian deliveries.

The Yelabuga facility, based on satellite imagery analysis and defector reports, appeared to employ over 1,000 workers across multiple shifts. The production process was surprisingly labor-intensive given the simple design. The foam and plywood construction required manual cutting and assembly. Electronic components needed hand-soldering. Each drone was individually tested before delivery. The process resembled aircraft manufacturing from the 1940s more than modern automated production.

This labor intensity had advantages. The production could scale rapidly by adding workers and shifts rather than requiring expensive automation equipment. Training times were short, weeks rather than months. And the dispersed, manual nature of production made the facilities harder to disrupt through precision strikes. Destroying a highly automated factory might cripple production; destroying a building full of workers and workbenches just meant moving operations elsewhere.

Quality control presented ongoing challenges. The Shahed design had inherent reliability issues, the simple construction and commercial-grade components meant failure rates were high even when manufacturing was done properly. Russian production, rushed and using improvised supply chains, experienced even higher failure rates. Ukrainian air defense reported that 15-20% of incoming Shaheds crashed or malfunctioned without being engaged, victims of engine failures, navigation errors, or structural problems.

But high failure rates mattered less when production volumes were high. If Russia could produce 500 Shaheds monthly and 100 failed in flight, 400 still reached Ukrainian airspace, more than enough to overwhelm air defenses and strike intended targets. The sheer volume compensated for individual unreliability.

Russian engineers worked continuously to improve production quality and enhance the design. By late 2023, newer Geran-2 variants showed measurably better reliability than early production models. The learning curve typical of any manufacturing program was compressing timeframes that normally took years into months of wartime urgency.

The Winter Campaign

The true scale of Russian Shahed production became evident in the winter campaign of 2023-2024. Beginning in November 2023 and intensifying through the coldest months, Russia launched waves of Shaheds against Ukrainian cities almost nightly. The targets were systematic: electrical infrastructure, heating plants, water pumping stations, the systems that made urban life possible during Ukrainian winters.

The campaign's arithmetic was brutal. Each Shahed cost Russia perhaps $50,000 to produce. Shooting down a Shahed with a Western air defense missile cost Ukraine potentially millions, depending on the system used. Russia could afford to trade ten Shaheds for one downed aircraft or destroyed infrastructure target and come out ahead economically. The cost asymmetry favored the attacker dramatically.

On typical nights, Ukraine faced 30-50 Shaheds approaching from multiple directions. On heavy attack nights, the numbers exceeded 100. The drones came in waves, timed to arrive at intervals that prevented Ukrainian air defenders from resting. A wave might arrive at midnight, another at 2 AM, another at 4 AM, forcing defenders to remain at heightened alert throughout the night.

Ukrainian air defenses adapted to the challenge with impressive effectiveness. Mobile fire groups, teams equipped with machine guns and MANPADS positioned around cities proved remarkably effective at shooting down the relatively slow Shaheds. Electronic warfare systems jammed or spoofed the drones' GPS navigation, causing them to crash in unpopulated areas. Radar-directed anti-aircraft artillery created kill zones that Shaheds had to traverse.

But volume remained Russia's advantage. Even with interception rates above 90%, if Russia launched 50 Shaheds and Ukraine shot down 47, three still got through. Three Shaheds striking power infrastructure could plunge portions of a city into darkness. Over weeks and months, the cumulative damage was severe.

The winter campaign revealed Russia's strategic calculus. They weren't trying to destroy Ukrainian air defenses directly that would have required more sophisticated weapons. Instead, they were grinding down Ukraine economically and psychologically. Each Shahed that Ukraine shot down represented ammunition expended, systems worn through use, and resources consumed. Each Shahed that got through caused damage that required repair, reconstruction, and restoration of services. Either way, Russia imposed costs.

Production Economics

The economics of Shahed production illuminated broader lessons about drone warfare and modern conflict. Traditional military procurement prioritized quality over quantity, better to have fewer exquisite weapons than many mediocre ones. The Shahed program inverted this calculation. Russia accepted high failure rates, crude construction, and limited capabilities in exchange for overwhelming numbers.

Component costs for a Shahed were estimated at $15,000-25,000, depending on supply chain efficiency. Manufacturing labor, facility costs, and overhead might add another $15,000-25,000. The total production cost likely ranged from $30,000-50,000 per unit, incredibly cheap for a weapon with 2,000-kilometer range and a 40-kilogram warhead.

At these costs, Russia could theoretically produce tens of thousands of Shaheds for the price of developing and building a single modern fighter aircraft. The strategic implications were profound. Traditional military thinking assumed air superiority required defeating enemy air forces through combat. The Shahed suggested an alternative: simply building so many long-range strike platforms that enemy defenses were overwhelmed through sheer volume.

The production economics also shaped Russian strategy. Because Shaheds were relatively cheap, Russia could afford to use them prodigally. Attacking targets of marginal military value was acceptable, the economic cost was sustainable. This enabled the terror bombing campaign against civilian infrastructure that more expensive weapons couldn't have sustained.

For Ukraine and its Western supporters, the Shahed presented a troubling challenge. How do you defend against weapons so cheap that the attacker can afford to build them faster than you can shoot them down? Traditional air defense doctrine assumed that interceptors were cheaper than targets, creating favorable exchange ratios for defenders. Shaheds inverted this relationship.

The only sustainable counter was either destroying production facilities, difficult when they were distributed and deep inside Russia or developing equally cheap counter-measures. Ukrainian mobile fire groups represented one solution: $100,000 in machine guns and training could shoot down millions of dollars worth of Shaheds. But even this required substantial investment in manning, training, and ammunition.

Variants and Evolution

As Russian Shahed production matured, new variants emerged addressing different operational requirements. The basic Geran-2 remained the workhorse, but specialized versions appeared throughout 2023 and 2024.

The Geran-2M featured enhanced range through more efficient engines and additional fuel capacity. Intelligence estimates suggested ranges exceeding 2,500 kilometers, allowing strikes on targets throughout western Ukraine from launch sites deep inside Russia. These versions sacrificed some warhead capacity for the fuel needed for extended missions.

Electronic warfare variants incorporated equipment to jam Ukrainian air defense radars or disrupt communications. These Shaheds typically flew alongside standard attack versions, providing electronic support for the strike packages. They weren't intended to survive, their countermeasures worked only for their own mission duration but they improved the chances that accompanying Shaheds would reach their targets.

Decoy variants were perhaps the most insidious. These Shaheds carried no warheads, instead incorporating radar reflectors and electronic emitters that made them appear threatening to air defense systems. Ukraine was forced to engage them with expensive interceptors or risk letting potentially live weapons through. Each decoy that drew fire saved a live warhead-carrying drone.

The variant proliferation showed Russian industrial adaptation in real-time. Unlike Western defense programs that required years to move from concept to production, Russian Shahed modifications appeared within months of operational needs being identified. The simple baseline design and flexible production process enabled rapid iteration.

By early 2024, Russian forces were launching coordinated strikes mixing multiple Shahed variants with cruise missiles, creating complex attack packages that stressed Ukrainian air defenses. The Shaheds served multiple roles: attacking targets directly, saturating defenses to enable missile penetration, jamming radars, and drawing fire away from more sophisticated weapons.

The Iranian Linkage Deepens

Throughout 2023 and 2024, Russia's relationship with Iran deepened beyond simple arms sales. Intelligence reporting suggested extensive technical cooperation, with Russian engineers advising on Iranian drone programs while Iranian specialists helped troubleshoot Russian production issues. The relationship became symbiotic, Iran gained access to combat feedback that informed their own programs, while Russia secured assured access to Iranian technical expertise and component supply chains.

There were reports, never fully confirmed, of Iranian drones being assembled in Russia using Iranian-supplied component kits before full technology transfer was complete. This would have allowed faster production scaling while Russian facilities were still being established. The boundary between Iranian and Russian production became deliberately murky, complicating Western intelligence assessments and sanctions enforcement.

The partnership raised troubling proliferation concerns. If Iranian technology could so readily transfer to Russia, where else might it go? North Korea maintained close ties with Iran. Syria operated under Iranian influence. Various non-state actors had relationships with Iran's Revolutionary Guards. The Shahed represented mature, proven technology that could spread throughout the networks of countries and groups operating outside Western-dominated international order.

Western officials struggled to respond effectively. Sanctioning Iranian entities involved in drone production had limited effect when those entities already operated under extensive sanctions. Interdicting component shipments required identifying them among millions of legitimate commercial electronics transfers. And destroying production facilities in Iran would have required military action that no Western government was prepared to authorize.

The Factory Continues

By late 2024, as this history was being compiled, Russian Shahed production showed no signs of slowing. Satellite imagery of the Yelabuga facility showed ongoing expansion. Attack rates against Ukraine remained high, some months seeing more Shaheds launched than in any previous period. The production network had achieved sustainability, weathering Western sanctions and Ukrainian strikes to maintain output.

Russian military doctrine had evolved to incorporate the Shaheds as a central element rather than a supplementary capability. Planning assumed availability of hundreds of Shaheds monthly. Tactics developed around the unique characteristics of cheap, expendable, long-range strike platforms. Russian forces had learned what Iran discovered years earlier: you didn't need exquisite technology when you could field overwhelming numbers.

The Shahed factory, both literal and metaphorical, represented a fundamental shift in how modern powers approached military production. The lesson wasn't lost on military planners worldwide. China, already a global leader in commercial drone production, undoubtedly noted how quickly Iran's relatively simple designs translated into strategic capability. Other nations with ambitions of regional power projection studied the model.

Russia had found a way to wage long-range strategic bombing without strategic bombers, to maintain pressure on Ukrainian infrastructure without depleting irreplaceable missile stockpiles, and to impose economic costs that Ukraine and its supporters struggled to counter. The Iranian partnership provided the blueprint; Russian industrial capacity, even constrained by sanctions, provided the scale.

As the war continued into 2025, the Shahed remained a persistent threat, a nightly terror for Ukrainian civilians, and a testament to how drone technology had reshaped modern warfare. The factory in Yelabuga and its counterparts would continue producing, the supply chains would continue smuggling components, and the waves of foam-and-plywood drones would continue launching into Ukrainian skies. Simple, cheap, and devastatingly effective the Shahed had found its place in military history.

Chapter 8

Ukrainian Innovation Surge

While Russia relied on Iranian technology and industrial-scale replication, Ukraine's response showcased a very different approach to drone warfare, one built on rapid innovation, distributed creativity, and the ability to iterate designs at a pace that traditional military procurement could never match. By late 2023, Ukraine had transformed from a country desperately importing commercial drones to a genuine drone innovation hub, producing platforms that in many ways surpassed what was available on the international market.

The Ukrainian innovation surge wasn't directed by central planners or driven by defense ministry requirements documents. It emerged organically from combat necessity, enabled by an ecosystem of small manufacturers, volunteer engineers, and frontline operators who communicated directly without bureaucratic intermediaries. The result was an explosion of designs, capabilities, and tactical applications that fundamentally changed what drones could accomplish on the battlefield.

The Pioneers Emerge

By spring 2023, certain Ukrainian drone manufacturers had distinguished themselves not just through production volume but through genuine innovation. These companies, often started by engineers, hobbyists, or entrepreneurs with no prior defense experience were solving problems that established defense contractors had spent years avoiding or addressing with expensive, complex solutions.

Escadrone, founded by a group of Kyiv-based engineers and drone racing enthusiasts, emerged as one of the early innovators. Their Pegasus platform represented a significant leap beyond the improvised FPV drones of 2022. The Pegasus featured a modular design that allowed rapid reconfiguration for different missions, reconnaissance, strike, electronic warfare, or relay operations. The frame was carbon fiber for strength and light weight, but designed for easy field repair with standardized components.

What set the Pegasus apart was its electronic warfare resistance. Escadrone's engineers had been iterating on anti-jamming systems since early 2023, testing different frequency hopping algorithms, power levels, and antenna configurations. By summer, the Pegasus could operate in electromagnetic environments that rendered most commercial drones useless. The system used dual-band transmission, automatically switching between 1.2 GHz and 5.8 GHz frequencies depending on jamming conditions, and incorporated directional antennas that focused signal strength toward the operator.

Taras, one of Escadrone's founding engineers, explained their development philosophy: "We weren't trying to build the perfect drone. We were trying to build something that worked today, that we could improve tomorrow, and that soldiers could actually use without weeks of training. Every week we'd get feedback from the front, this failed, that worked great, we need this capability. Within days, we'd have a modified design ready for testing. Try that with traditional defense contractors."

The Pegasus wasn't cheap by FPV standards, around $800-1,000 per unit but it delivered dramatically better performance than the $300 improvised alternatives. More importantly, it was designed for Ukrainian operational conditions: cold weather performance, resistance to Russian jamming, easy maintenance with limited tools, and compatibility with locally-available components.

Vyriy and the Reconnaissance Revolution

While Escadrone focused on strike platforms, Vyriy Drone took a different path, specializing in reconnaissance systems. Their Molfar platform, named after Ukrainian mystics who could see hidden things represented a significant advance in tactical intelligence gathering.

The Molfar was a fixed-wing design, dramatically different from the quadcopter platforms dominating the battlefield. Fixed-wing drones were more efficient for long-duration missions, capable of staying airborne for hours while quadcopters were limited to 20-30 minutes. The Molfar could loiter over target areas for extended periods, providing continuous surveillance that tactical commanders desperately needed.

But the real innovation was in the sensor package and data transmission. Vyriy equipped the Molfar with stabilized cameras that could zoom significantly while maintaining image quality, thermal imaging for night operations, and laser rangefinders for precise targeting data. The drone could identify targets at ranges where it was nearly invisible from the ground, providing surveillance without alerting enemy forces.

The data transmission system was equally sophisticated. Rather than streaming continuous video, which consumed enormous bandwidth and was vulnerable to interception the Molfar used intelligent compression and selective transmission. The onboard processor analyzed the video feed, identified significant changes or potential targets, and transmitted only relevant imagery. This reduced bandwidth requirements by 80-90%, making the drone much harder to jam or intercept.

By October 2023, Vyriy was producing several hundred Molfar platforms monthly, and the drones had become essential tools for Ukrainian artillery units. The typical workflow was elegant in its efficiency: a Molfar would identify Russian positions and relay GPS coordinates directly to fire control computers. Ukrainian artillery would fire. The Molfar would provide immediate battle damage assessment and spot corrections for follow-up strikes. The entire cycle from target identification to engagement took minutes rather than the hours traditional reconnaissance required.

Colonel Andriy, whose artillery brigade was among the first to operate with dedicated Molfar support, described the impact: "Before, we were essentially firing blind half the time. We'd get reports of enemy positions, maybe some coordinates that were hours old, and we'd fire hoping to hit something. With the Molfar overhead, we could see exactly what we were shooting at, watch our rounds impact, and adjust fire in real-time. Our effectiveness probably tripled."

TAF Drones: Scaling Innovation

While smaller manufacturers like Escadrone and Vyriy pushed boundaries on specific platforms, TAF Drones emerged as Ukraine's first truly large-scale drone manufacturer. Founded in early 2023 by a consortium of investors and defense industry veterans, TAF represented an attempt to bring industrial discipline to the chaotic world of Ukrainian drone manufacturing.

TAF's approach was different from the garage workshops. They established a proper factory facility in central Ukraine with assembly lines, quality control processes, and standardized components. They hired engineers away from automotive and electronics companies, offering salaries that could compete with private sector alternatives. They invested in tooling and equipment that allowed precision manufacturing rather than hand-assembly.

The bet was that systematic manufacturing could deliver better drones at competitive prices once production reached sufficient scale. It was a risky proposition—the upfront investment was substantial, and the distributed garage model seemed to be working adequately for most applications. But TAF's founders believed that as the war continued, demand would favor manufacturers who could guarantee quality and quantity simultaneously.

By summer 2023, TAF was producing approximately 2,000 drones monthly across multiple platform types. Their Strike-1 FPV kamikaze drone became one of the most widely used Ukrainian platforms. It wasn't dramatically superior to competitors, but it was consistent, batches of 100 Strike-1s would have similar performance characteristics, making them predictable for operators and maintainers.

TAF also invested heavily in training. They established a school for drone operators, teaching not just flight skills but maintenance, troubleshooting, and tactical employment. Graduates received certification and preference for receiving TAF equipment. This created a virtuous cycle: better-trained operators provided better feedback, which informed better designs, which attracted more customers, which funded more training.

The company's growth through 2023 was explosive. By October, monthly production had reached 5,000 units. By December, over 8,000. TAF had become Ukraine's largest single drone manufacturer, and their products were appearing across virtually every sector of the front. The company's valuation, though never publicly disclosed, was rumored to exceed $500 million, remarkable for a company that hadn't existed eighteen months earlier.

The Innovation Ecosystem

What made Ukrainian drone innovation so effective wasn't any single company or platform, it was the ecosystem that allowed rapid knowledge transfer and collaborative development. This ecosystem operated through multiple channels, most of them informal and organic rather than officially organized.

Telegram channels became crucial information exchanges. Drone manufacturers, volunteer organizations, and frontline operators all participated in various channels where combat footage, technical discussions, and tactical innovations were shared freely. A clever modification developed by one workshop would be discussed, analyzed, and adopted by dozens of others within days. Problems encountered in the field would generate immediate troubleshooting discussions involving engineers who had never met in person.

The competitive dynamic was unusual. Manufacturers competed for customers and funding, but they also collaborated freely on shared challenges. When Russian jamming intensified in a particular frequency band, multiple manufacturers would share findings about effective countermeasures. When a new type of Russian electronic warfare system was identified, the community would collectively work on solutions. The imperative of Ukrainian survival created a collaborative urgency that transcended commercial competition.

Physical meetups, though less frequent, were equally valuable. Manufacturers, volunteers, and military representatives would gather in Kyiv, Lviv, or other cities for informal conferences where drones were displayed, capabilities demonstrated, and relationships built. These weren't formal defense exhibitions, they were more like drone enthusiast gatherings, except the stakes were existential and the customers were combat units fighting for national survival.

Foreign technical assistance accelerated innovation. Ukrainian manufacturers maintained relationships with drone racing communities, commercial drone engineers, and defense contractors worldwide. Technical discussions happened through encrypted channels, with foreign experts providing advice on everything from flight controller programming to composite materials. Some foreign volunteers even traveled to Ukraine to work directly with manufacturers, contributing specialized expertise that would have taken months or years to develop domestically.

The innovation ecosystem also benefited from Ukraine's broader technical talent pool. The country had a substantial IT sector, with tens of thousands of software engineers, many of whom had worked for international tech companies. When these professionals turned their attention to military drones, they brought sophisticated software engineering practices, agile development methodologies, and modern programming capabilities. Ukrainian drones began incorporating features such as autonomous navigation, computer vision and swarm coordination that required software expertise well beyond traditional defense contractors.

The Production Race

By October 2023, the production competition between Ukraine and Russia had reached remarkable intensity. Intelligence assessments suggested both countries were producing at comparable rates, each manufacturing thousands of drones monthly. The specific numbers remained classified, but the rough parity was evident in attack and defense patterns across the front.

This parity represented a stunning achievement for Ukraine. Russia had significant advantages: a larger industrial base, more manufacturing capacity, easier access to components through less-scrutinized supply chains, and Iranian technical support. That Ukraine matched Russian production despite these disadvantages demonstrated the effectiveness of the distributed, innovation-focused model.

The production race drove continuous improvement on both sides. Neither country could afford to rest on existing designs when the opponent was constantly innovating. Russian drones captured by Ukrainian forces would be disassembled, analyzed, and any useful features would appear in Ukrainian designs within weeks. Ukrainian drones recovered by Russian forces underwent similar analysis, driving Russian improvements.

This created a technological ratchet effect, capabilities that appeared cutting-edge in January would be standard by June and obsolete by

December. The pace of innovation exceeded anything in peacetime defense development, where programs typically spanned decades from concept to fielding.

The production numbers were staggering by any historical standard. In World War II, the Soviet Union produced roughly 40,000 combat aircraft annually at peak production. In 2023, Ukraine alone was approaching similar numbers for drones, admittedly much simpler and cheaper systems, but still remarkable given the country's smaller population and industrial base.

The sustainability of this production race became a key strategic question. Both countries were burning through drones at rates that would have seemed insane a year earlier. Hundreds were being lost daily, destroyed by enemy fire, crashed due to mechanical failure, disabled by electronic warfare, or simply expended in successful attacks. Could either side maintain these production rates indefinitely? What happens when component supplies are tightened or manufacturing capacity saturated?

Specialized Platforms Proliferate

As the innovation surge continued, Ukrainian manufacturers began developing increasingly specialized platforms for specific mission types. The era of general-purpose drones was giving way to a diverse ecosystem of purpose-built systems.

Heavy-lift quadcopters emerged for dropping anti-tank mines and larger munitions. These platforms could carry payloads of 5-10 kilograms, allowing them to deliver weapons that smaller FPV drones couldn't manage. They were slower and more vulnerable than FPV racers but filled a crucial capability gap for certain targets and tactics.

Long-range strike drones, essentially Ukrainian equivalents to the Shahed began appearing in larger numbers. Platforms with 500+ kilometer ranges allowed strikes on Russian logistics hubs, command posts, and infrastructure deep behind the lines. These weren't as sophisticated as cruise missiles, but they were vastly cheaper and could be produced domestically.

Electronic warfare drones carried specialized equipment for jamming Russian communications, disrupting drone operations, or providing cover for Ukrainian forces. Some versions could detect and locate Russian electronic warfare systems, providing targeting data for artillery or long-range strikes.

Interceptor drones represented perhaps the most innovative development, drones designed to hunt other drones. These platforms carried sensors optimized for detecting other UAVs and could engage them either kinetically or through electronic attack. The concept of drone-versus-drone combat moved from theory to operational reality.

Maritime drones, though technically outside the aerial drone category, showed similar innovation. Ukrainian manufacturers developed remote-controlled surface vessels capable of attacking Russian naval assets, contributing to Ukraine's success in contesting Black Sea control despite having virtually no conventional navy.

Quality Control Challenges

The rapid production scaling and distributed manufacturing model created inevitable quality control challenges. Not every workshop maintained rigorous standards, and the pressure to maximize production sometimes trumped reliability concerns.

Frontline units reported significant variation in drone quality between manufacturers. Some workshops produced drones that failed at unacceptably high rates, electronics that burned out, motors that seized, frames that cracked under stress. Soldiers learned through experience which manufacturers were reliable and which to avoid, creating informal reputation systems that rewarded quality and punished carelessness.

The problem was particularly acute for electronic components. Counterfeit parts infiltrated supply chains, often indistinguishable from genuine components until they failed under operational stress. Flight controllers that worked fine in testing would crash when exposed to combat vibrations or temperature extremes. Batteries would deliver far less capacity than advertised, leaving drones with insufficient endurance for their missions.

TAF Drones and other larger manufacturers responded by implementing more rigorous incoming inspection and testing protocols. They established relationships with trusted component suppliers and paid premiums for verified genuine parts. They implemented burn-in testing where drones were operated for extended periods before delivery to identify infant mortality failures. These measures increased costs and reduced production rates slightly, but they also dramatically improved reliability.

The Ukrainian military attempted to establish some standardization and certification, but the effort struggled to keep pace with the rapid evolution of designs and manufacturers. By the time a drone model completed evaluation and certification, newer versions with different characteristics were already in production. The certification process risked becoming an obstacle to innovation rather than a guarantee of quality.

Ultimately, the quality control problem was managed through market mechanisms rather than centralized control. Combat units had budgets for drone procurement and chose manufacturers based on performance and reputation. Manufacturers whose products failed excessively lost customers and went out of business. Those whose products performed reliably gained market share and expanded. It was Darwinian capitalism applied to military procurement, and while imperfect, it drove continuous quality improvement.

The International Dimension

Ukraine's drone innovation attracted significant international attention. Foreign military observers studied Ukrainian tactics and technologies, recognizing that Ukraine was writing the doctrine for drone warfare in real-time. NATO members sent delegations to learn from Ukrainian experience. Defense contractors studied Ukrainian innovations, recognizing that garage workshops in Kyiv and Dnipro were solving problems that Western defense giants had struggled with for years.

Some international support was material, foreign governments and organizations funded Ukrainian drone production through various mechanisms. Other support was technical, Western engineers, software developers, and drone experts provided advice and expertise. The relationship was symbiotic: Ukraine gained resources and knowledge while international partners gained insights into the future of warfare.

There were also emerging commercial opportunities. Several Ukrainian manufacturers began receiving inquiries from foreign buyers interested in their platforms. While wartime production focused entirely on domestic military needs, there was growing recognition that Ukrainian drone companies could become significant defense exporters once the conflict ended. The combination of combat-proven designs, low costs, and innovation capabilities made Ukrainian drones potentially very competitive in international markets.

October 2023: A Milestone

By October 2023, the Ukrainian drone innovation surge had reached a remarkable milestone: production parity with Russia. Despite Russia's advantages in industrial capacity and Iranian support, Ukraine's distributed innovation model had proven capable of matching their output. Both sides were producing thousands of drones monthly, consuming them in daily combat, and continuously iterating on designs.

This parity represented a strategic achievement that few had predicted. In February 2022, Ukraine had virtually no domestic drone industry. Twenty months later, they were matching production with a country whose economy was fifteen times larger. The transformation validated the innovation-focused, distributed manufacturing model and demonstrated that in modern warfare, adaptability and creativity could compensate for material disadvantages.

But parity wasn't victory. The production race would continue, consumption rates would increase, and both sides would push toward even higher output. The innovation surge had positioned Ukraine to compete, but the war's outcome would depend on who could sustain the pace longer and who could translate numerical output into battlefield effectiveness.

The race was far from over. If anything, it was just beginning.

PART III: THE MASS DRONE ERA (2024)

Chapter 9

Growing Bigger, Flying Farther

The FPV drone that Maksym held in his hands in January 2024 looked nothing like the racing quadcopter he'd flown a year earlier. Where once sat a sleek seven-inch frame designed for agility and speed, now sprawled a bulky thirteen-inch monster, its carbon fiber arms thick as his thumb, motors the size of shot glasses humming with barely contained power. The transformation wasn't aesthetic, it was existential. The Ukrainian drone war was growing up, and growing bigger in every measurable way.

"We started with what we had," Maksym explained, his fingers instinctively checking the connections on the drone's flight controller. "Racing drones. Toys, really. You could carry maybe 400 grams of explosive. Sometimes just a grenade taped to the bottom. Now?" He gestured to the warhead mounted beneath the frame, a shaped charge weighing nearly two kilograms. "Now we can kill anything on the battlefield."

From Racing Drones to Flying Artillery

The evolution of the FPV drone from 2023 to 2024 represented one of the most rapid technological adaptations in military history. The standard seven-inch racing drone frame that dominated early 2023 warfare had been purpose-built for speed and maneuverability in drone racing competitions. Pilots would thread these machines through gates and around obstacles at speeds exceeding 150 kilometers per hour, performing barrel rolls and split-second turns that would grey out a human pilot. These same characteristics made them devastatingly effective against stationary targets, trenches, vehicles, and supply dumps in the first phase of drone warfare.

But by late 2023, both Ukrainian and Russian forces had identified critical limitations. The seven-inch frame's lightweight construction meant payload capacity topped out at around 500 grams. While sufficient for anti-personnel attacks or destroying light vehicles, these drones struggled against hardened targets. Tanks, armored personnel carriers, and fortified positions required heavier warheads to penetrate. Equally problematic, the standard racing drone battery configuration provided only 5-7 minutes of flight time, limiting operational range to perhaps three to four kilometers at best.

The solution emerged not from central planning but from the distributed network of Ukrainian drone manufacturers operating in apartments, garages, and small workshops across the country. Engineers and pilots began experimenting with larger frame sizes throughout late 2023, testing ten-inch, twelve-inch, and eventually thirteen-inch configurations. Each increment brought trade-offs. Larger frames meant more powerful motors, larger propellers, bigger batteries and exponentially greater payload capacity.

By January 2024, the thirteen-inch frame had emerged as the new standard. These platforms could carry warheads weighing two to three kilograms while maintaining flight times of 15-20 minutes. The operational implications were immediate and dramatic. Where a seven-inch drone might damage a vehicle, a thirteen-inch drone could destroy it. Where smaller drones bounced off tank armor, the new generation could penetrate and kill.

"The first time we hit a Russian T-72 with one of the big drones, I didn't believe what I was seeing," recalled Oleksandr, a drone company operator in the Donetsk region. "The explosion wasn't just bigger, it was purposeful. The shaped charge we mounted could punch through the top armor. One drone, one tank. That changed everything."

The Payload Revolution

The increased payload capacity of larger FPV drones catalyzed a parallel revolution in warhead design and mission flexibility. Ukrainian engineers, many with backgrounds in mechanical engineering, chemistry, or simply possessing autodidactic determination, began developing specialized munitions optimized for drone delivery.

Anti-tank warheads topped the priority list. Drawing on principles dating back to World War II's shaped charge technology, Ukrainian workshops produced copper-cone penetrators that could defeat armor plate up to 200 millimeters thick. Mounted on thirteen-inch drones, these warheads transformed FPV platforms from harassment weapons into legitimate tank killers. Russian armor, designed to withstand frontal attacks from anti-tank missiles and main gun rounds, proved vulnerable to top-attack strikes against thinner upper hull and turret armor.

Thermobaric warheads emerged as another innovation, particularly effective against personnel in trenches and fortified positions. These fuel-air explosives created devastating overpressure waves that could clear bunkers and dugouts without requiring direct hits. A drone carrying a thermobaric charge could dive into a trench entrance and eliminate an entire squad in the resulting blast.

Fragmentation munitions, essentially enhanced grenades with pre-fragmented casings, proved ideal for attacking exposed infantry and soft-skinned vehicles. The larger drones could carry multiple submunitions, turning a single FPV platform into a flying cluster bomb capable of devastating areas hundreds of square meters in size.

But payload capacity meant more than just bigger bombs. Ukrainian forces began equipping FPV drones with additional sensors, communications equipment, and electronic warfare packages. Drones carried signal repeaters that extended the range of other drones operating beyond line of sight. Some mounted small thermal cameras providing targeting capability in darkness or through smoke. Others carried electronic warfare modules that could jam Russian communications or trigger improvised explosive devices from safe distances.

The thirteen-inch platform's versatility extended to non-kinetic missions as well. Medical supply delivery became routine, with drones ferrying blood, antibiotics, and tourniquets to isolated forward positions under fire. Commanders used heavy-lift drones to resupply ammunition to units cut off by artillery fire. One innovative battalion even used modified FPV drones to carry fresh batteries to reconnaissance teams operating behind Russian lines, extending missions that otherwise would have required dangerous resupply runs.

Extended Range: The Strategic Dimension

If payload capacity represented the vertical dimension of drone evolution, range represented the horizontal. The seven-inch drones of 2023 could barely reach targets five kilometers away before battery depletion forced their return or more commonly, their kamikaze mission concluded with impact. The thirteen-inch platforms, with their larger batteries and more efficient power systems, pushed operational ranges to 15-20 kilometers, fundamentally altering the strategic calculus of drone warfare.

This extended range meant Ukrainian forces could strike targets previously considered safe. Russian artillery positions that had operated with impunity 10 kilometers behind the front lines now found themselves within the engagement envelope of FPV swarms. Command posts, logistics nodes, and reserve assembly areas all became vulnerable. The Russians faced an impossible choice: either push critical assets even farther back, degrading their effectiveness and complicating their own logistics, or accept escalating losses to drone strikes.

The range extension also enabled deep surveillance operations. Reconnaissance-configured drones could penetrate far behind enemy lines, loiter over key locations, and return with actionable intelligence. Unlike satellite imagery or signals intelligence, these drones provided real-time video feeds that commanders could use for immediate targeting decisions.

A Ukrainian battery commander could watch a drone feed of Russian forces assembling for an attack and have artillery rounds impacting those positions within minutes.

Ukrainian engineers achieved these range improvements through multiple technical innovations. High-capacity lithium polymer batteries, often the same cells used in electric vehicles, replaced standard racing drone batteries. More efficient motor and propeller combinations reduced power consumption while maintaining thrust. Lightweight carbon fiber frames shed unnecessary weight, allowing more energy to be devoted to propulsion rather than lifting the drone itself.

Some manufacturers pursued even more ambitious range extension programs. Modified FPV drones with oversized batteries and streamlined aerodynamics achieved ranges exceeding 30 kilometers, though at the cost of reduced maneuverability and longer flight times that increased vulnerability to electronic warfare. These ultra-long-range platforms occupied a niche between tactical FPV drones and strategic loitering munitions like the Shahed-136, capable of striking targets well behind the front lines but with the precision and real-time control that waypoint-programmed drones lacked.

The $2.47 Billion Commitment

On a gray morning in March 2024, Ukrainian President Volodymyr Zelensky stood before assembled defense industry representatives and announced a figure that captured the scale of Ukraine's commitment to drone warfare: the government had contracted for the manufacture of 1.6 million unmanned aerial vehicles at a cost of $2.47 billion. The announcement sent shockwaves through military analysis communities worldwide.

To understand the magnitude of this commitment, consider that 1.6 million drones represented more UAVs than existed in the entire world's military inventories combined at the start of the war. The United States military, the most technologically advanced force on Earth, operated fewer than 10,000 drones of all types in February 2022. Ukraine was preparing to produce more drones in a single year than most nations would acquire in a decade.

The per-unit cost revealed equally striking data: at $2.47 billion for 1.6 million units, Ukraine was paying an average of roughly $1,540 per drone. This figure masked significant variation, simple FPV kamikaze drones cost as little as $400-500 each, while sophisticated reconnaissance platforms with thermal cameras and encrypted communications cost $5,000 or more. But the average spoke to the democratization of military airpower that drones represented. For the cost of a single American Reaper drone ($32 million), Ukraine could field over 20,000 FPV drones.

The contract represented more than just procurement, it was industrial policy. Rather than consolidating production with a few large defense contractors, Ukraine distributed orders across dozens of manufacturers. TAF Drones, the largest producer, received contracts for hundreds of thousands of units. But smaller manufacturers like Escadrone, Wild Hornets, and Vyriy also secured substantial orders. Even individual combat units that had established their own mini-factories received official contracts to produce drones for wider distribution.

This distributed approach served multiple purposes. It made the production system resistant to Russian strikes, destroying one factory wouldn't cripple the entire program. It fostered competition and innovation, as manufacturers raced to develop better, cheaper, more capable drones to secure future contracts. And it built an industrial ecosystem with deep roots throughout Ukrainian society, creating skills and knowledge that would persist long after the war ended.

The $2.47 billion also represented something psychological and political. It was Ukraine doubling down on the technology that had proven most effective in combat. It was a message to Russian forces that the drone swarms would not diminish but intensify. And it was a signal to Western allies that Ukraine was taking responsibility for its own defense in the domains where it had proven most innovative.

Production at Scale: The Factory Floor

To meet the astronomical demand of 1.6 million drones, Ukrainian manufacturing had to industrialize processes that had begun in garages and workshops. The transformation of TAF Drones illustrated this evolution. In early 2023, TAF operated from a small facility in Kyiv where a dozen workers hand-assembled perhaps 50 drones daily. By early 2024, the company had expanded to multiple facilities employing hundreds of workers producing over 1,000 drones per day, more than 30,000 per month.

Walking through a TAF facility in spring 2024 felt like entering a different world from the guerrilla manufacturing of just months earlier. Assembly lines stretched across warehouse floors, each station specialized for particular tasks. At the first stations, workers mounted motors to carbon fiber arms, applying precise torque specifications to ensure reliability under combat stress. Down the line, others soldered flight controllers, testing each connection with multimeters to catch defects before final assembly.

Quality control represented a constant challenge. In the early days, drone builders simply assembled units and shipped them to the front, letting combat be the ultimate test. But as production scaled, systematic failures became intolerable. A batch of 1,000 drones with faulty motor controllers wasn't just a production setback, it was a military disaster that could cost lives and battlefield success. TAF and other major manufacturers instituted testing regimes that included flight testing every tenth unit, vibration testing for structural integrity, and electrical testing of all components.

The supply chain complexity of drone production at this scale proved formidable. A single FPV drone contained hundreds of components: motors, electronic speed controllers, flight controllers, receivers, antennas, batteries, frames, propellers, wiring, connectors, cameras, and video transmitters. Each component had to be sourced, inventoried, and delivered to assembly lines in precise synchronization. A shortage of a single type of motor could halt production across an entire facility.

Chinese components dominated the supply chain, despite China's official restrictions on military drone exports to conflict zones. Motors manufactured by T-Motor and electronics from manufacturers like Radiomaster and JHEMCU found their way to Ukraine through circuitous routes involving third countries and intermediary buyers. Some Ukrainian manufacturers established subsidiary companies in neighboring countries to obscure the ultimate destination of components. The cat-and-mouse game of supply chain logistics became as important as the tactical cat-and-mouse of electronic warfare on the battlefield.

Labor emerged as another bottleneck. Assembling FPV drones required skilled workers who understood electronics, mechanics, and the specific requirements of combat drones. TAF and other companies established training programs, often recruiting from technical universities and the hobbyist drone racing community. Workers learned to solder, test, assemble, and troubleshoot drones through intensive multi-week courses. The best graduates could assemble a complete FPV drone from components in under 45 minutes while maintaining the precision necessary for reliable operation.

Wages in the drone industry reflected both the skilled nature of the work and the urgency of wartime production. Assembly line workers earned substantially more than typical manufacturing wages, often $1,000-1,500 per month, significant in an economy strained by war. Experienced workers who could train others or troubleshoot complex problems commanded even higher compensation. The drone industry became one of the few growth sectors in Ukraine's wartime economy, drawing workers from other industries and providing employment in cities far from the front lines.

The October 2023 Milestone

By October 2023, even before the massive 1.6 million drone contract, Ukrainian production had reached an inflection point. Multiple sources confirmed that Ukraine was manufacturing thousands of drones per month, with some estimates suggesting 5,000-7,000 units monthly across all manufacturers. More significantly, this production rate roughly matched Russian output, despite Russia's vastly larger economy and industrial base.

This parity represented a remarkable achievement. Russia could draw on Soviet-era military-industrial infrastructure, deeper financial resources, and a larger pool of engineers and workers. Yet Ukraine's distributed production model, fueled by volunteer initiatives, crowdfunding, and government contracts, had matched the output of its larger adversary. The achievement spoke to both Ukrainian ingenuity and Russian structural disadvantages such as Moscow's centralized, corruption-prone defense industry struggling to adapt to the rapid innovation cycles that drone warfare demanded.

The production parity also suggested that the war had entered a new phase: industrial attrition. Both sides could generate drones in large quantities, meaning success would depend not just on having drones, but on using them more effectively, protecting them from electronic warfare, and integrating them into combined arms operations. The romanticized vision of scrappy Ukrainian volunteers outproducing Russian factories had given way to a grimmer reality, both nations were locked in an industrial competition that would consume vast resources and countless drones in an ongoing war of material.

Evolution Never Stops

Even as thirteen-inch drones became standard in early 2024, Ukrainian engineers were already designing the next generation. Experimental platforms featuring fifteen-inch and even seventeen-inch frames emerged from workshops, testing the limits of what could still be considered an FPV drone versus entering the realm of larger tactical UAVs. These giants could carry payloads exceeding five kilograms and achieve ranges of 30-40 kilometers, blurring the line between kamikaze drone and loitering munition.

Hybrid designs appeared, combining FPV maneuverability with fixed-wing efficiency. These drones could fly like conventional aircraft for most of their mission, conserving battery power, then switch to quadcopter mode for final approach and precision strike. The complexity was greater, but so was the operational flexibility, a hybrid drone could loiter over a target area for extended periods, waiting for the ideal target to appear.

Specialized variants proliferated as different units identified specific needs. Naval drones adapted from FPV platforms began targeting Russian patrol boats in the Black Sea. Urban warfare variants featured quieter motors and propellers for operations in cities where noise discipline mattered. Winter variants incorporated heated batteries and cold-weather electronics to maintain performance in sub-zero temperatures that caused standard batteries to fail.

The innovation cycle had compressed to weeks instead of months. A front-line unit would identify a need, drones that could operate in heavy electronic warfare environments, for instance. They'd communicate with their manufacturer, often a small company they'd worked with for months. Engineers would design a solution, test it, and have modified drones at the front within two to three weeks. This rapid iteration, impossible in traditional military procurement systems that required years of testing and certification, gave Ukraine a qualitative edge that compensated for any quantitative shortfalls.

The Human Cost of Production

Behind the statistics and technical specifications lay human beings working under extraordinary pressure. Drone production facilities, while far from the front lines, weren't safe. Russian missiles regularly targeted Ukrainian defense industry sites, and drone factories represented prime targets. Workers assembled drones knowing that at any moment, air raid sirens might send them scrambling to shelters, or worse, that their facility could be hit before warning arrived.

The psychological weight of the work was profound. Every worker knew that the drones they assembled would kill Russian soldiers and that Russian drones, perhaps assembled in similar facilities by similar workers, would kill Ukrainians. Some workers found meaning in this, viewing their labor as a direct contribution to national defense. Others struggled with the moral implications, building weapons they'd never see used but knowing their handiwork was delivering death.

Yulia, a 28-year-old engineer at a Lviv drone manufacturer, described the tension: "I studied aerospace engineering because I loved the idea of flight, of exploration. Now I spend my days calculating the best way to attach explosives to drones. It's not what I imagined. But it's necessary. Every drone we build that works perfectly might save Ukrainian lives. That has to be enough."

The pace was relentless. Twelve-hour shifts were common, seven-day work weeks standard during surge production periods. Quality couldn't be sacrificed for speed, a faulty drone wasn't just a waste of money but could mean a failed mission or Ukrainian casualties if forces relied on drones that didn't perform. The pressure to balance speed and quality, quantity and reliability, wore on workers and management alike.

Looking Forward

As 2024 progressed, the trajectory was clear: drones were getting bigger, more capable, more numerous, and more central to warfare. The thirteen-inch platforms of early 2024 would be standard by summer and possibly obsolete by year's end as even larger, more sophisticated systems entered service. Ukraine's contracted 1.6 million drones for 2024 represented not a ceiling but a floor, actual production might exceed that number as manufacturers expanded capacity and new producers entered the market.

The evolution from seven-inch racing drones to thirteen-inch strike platforms had taken barely a year. What would another year bring? Fifteen-inch frames were already in testing. Hybrid designs combining multiple propulsion methods showed promise. Artificial intelligence and autonomous systems loomed on the horizon. The drone war was accelerating, and no one could predict where the technology would go next.

What was certain was that Ukraine had committed fully to this new form of warfare. The $2.47 billion contract, the hundreds of manufacturers, the thousands of workers, the millions of drones, all of it represented an unprecedented bet that small, cheap, numerous unmanned systems would prove decisive in modern warfare. The bet was made not from theory but from experience, from two years of combat that had proven drones' effectiveness again and again.

Maksym finished his pre-flight check of the thirteen-inch drone, its motors spinning up with a deep, purposeful hum unlike the high-pitched whine of the racing drones he'd flown in another life. The warhead secured beneath it represented lethal potential that would have required artillery or missiles in previous wars. Now it could be delivered by a device built in an apartment workshop for $800.

"We're flying farther and hitting harder," he said, watching the drone lift off and accelerate toward the horizon. "And we're just getting started."

Chapter 10

The Fiber-Optic Solution

The spool looked innocuous sitting on the workbench, just a few kilometers of hair-thin fiber-optic cable wound around a plastic cylinder, the kind of material that might connect a home to high-speed internet. But Dmitry Koval, a Ukrainian drone engineer in Dnipro, knew this unassuming spool represented the most significant breakthrough in FPV drone technology since the war began. It was March 2024, and Russian forces had just deployed the first fiber-optic guided drones in combat. Within weeks, Ukraine would not only replicate the technology but refine it, launching both sides into a new era of jam-proof warfare.

"When we first saw footage of a Russian fiber-optic drone," Dmitry recalled, "my immediate reaction was 'of course.' It was so obvious, I couldn't believe we hadn't thought of it first. Electronic warfare had made radio-controlled drones nearly useless in some sectors. The Russians were jamming everything. We were jamming them. Drones were falling from the sky before reaching targets. We needed something that couldn't be jammed, and the answer was literally as old as telecommunications itself, a physical wire."

The Jamming Crisis

To understand why fiber-optic guidance represented such a breakthrough, one must first grasp the electronic warfare nightmare that had evolved by early 2024. The electromagnetic spectrum over the Ukrainian battlefield had become the most contested invisible territory in human history. Both sides deployed increasingly sophisticated jamming systems that turned the airwaves into static chaos, creating dead zones where radio-controlled drones simply couldn't function.

Russian electronic warfare systems like the Krasukha-4 and Borisoglebsk-2 could blanket entire sectors with jamming across multiple frequency bands simultaneously. These systems didn't just disrupt drone control signals, they attacked GPS navigation, video transmission, and even the telemetry data that pilots used to understand their drone's status. A drone entering heavy jamming might lose its video feed first, leaving the pilot blind. Then control signals would fail, leaving the aircraft unresponsive. Finally, GPS would spoof or disappear entirely, leaving the drone's flight controller confused about its position and altitude.

Ukrainian forces deployed their own electronic warfare capabilities with equal aggression. Systems purchased from Western allies or developed domestically created jamming bubbles around key positions, headquarters, and logistics nodes. The result was an electromagnetic hell where drones that cost hundreds of dollars in hardware and represented hours of skilled labor would fly straight into invisible walls of jamming and tumble from the sky, their missions unfulfilled, their operators cursing in frustration.

The statistics told the story. By late 2023, some units reported mission failure rates exceeding 70% in heavily jammed areas. A drone company might launch ten FPV strikes against Russian positions and watch seven crash uselessly before reaching their targets, victims of jamming they couldn't overcome. The attrition wasn't just drones, it was morale, resources, and tactical opportunity. Electronic warfare had created a defensive advantage that threatened to make drone warfare obsolete before it fully matured.

Engineers on both sides attacked the problem from multiple angles. Frequency-hopping spread spectrum systems tried to outmaneuver jammers by constantly changing transmission frequencies. Higher-power transmitters attempted to punch through jamming with brute force. Directional antennas focused signals more precisely, reducing vulnerability to area jamming. Some manufacturers experimented with extremely low frequency transmissions that were harder to jam but carried less data.

None of these solutions proved decisive. Frequency-hopping could be countered by wideband jamming that covered entire frequency ranges simultaneously. Higher power transmitters made ground stations more vulnerable to anti-radiation missiles that homed in on electronic emissions. Directional antennas required maintaining precise alignment between pilot and drone, difficult during high-speed maneuvering. Every electronic countermeasure spawned an electronic counter-countermeasure in an endless spiral of technological one-upmanship.

The breakthrough would come not from more sophisticated electronics, but from abandoning electronics entirely for the critical control link.

Russia Deploys First

Intelligence reports in March 2024 brought disturbing news to Ukrainian drone units: Russian FPV drones were suddenly achieving unprecedented success rates in heavily jammed areas. Drones that should have been falling from the sky were boring through Ukrainian electronic warfare systems as if they didn't exist, striking with precision against Ukrainian armor and positions. Initially, Ukrainian analysts suspected the Russians had developed new encryption or frequency-hopping algorithms. The truth, when discovered through examination of shot-down Russian drones, was simultaneously simpler and more elegant.

The Russians had gone analog. Sort of.

The captured Russian drone revealed a fiber-optic cable spooled in the tail section, paying out behind the aircraft as it flew toward its target. The cable, barely thicker than a human hair, carried light pulses encoding the pilot's control inputs and the drone's video feed. Light traveling through glass fiber was immune to radio frequency jamming. The physics were inescapable, you could fill the air with electromagnetic noise, but it wouldn't affect photons traveling through a strand of glass.

The Russian implementation was crude but effective. The cable spool added weight, limiting range and reducing maneuverability. The mechanical payout system sometimes jammed, causing the drone to crash. The cable itself would occasionally snag on obstacles or break under stress. But when it worked, it worked perfectly. Russian pilots could control their drones with zero latency and complete immunity to jamming, flying with confidence through electronic warfare environments that had been impassable weeks earlier.

The tactical advantages manifested immediately. Russian fiber-optic drones began systematically targeting Ukrainian electronic warfare systems, striking the very installations that should have been protecting forces from drone attacks. Jamming equipment that had provided an invisible shield suddenly became worthless against this new threat. Ukrainian commanders watched with growing alarm as their defensive advantages evaporated.

Russian military bloggers and Telegram channels celebrated the new technology as a "game-changer." Videos showed fiber-optic drones maneuvering through heavy fire, their pilots maintaining perfect control despite massive electronic warfare efforts. One particularly dramatic video captured a Russian fiber-optic drone diving into a Ukrainian M113 armored personnel carrier, the cable visible in the final frames as a thin line stretching back toward the horizon.

Ukraine's Rapid Response

Ukraine's drone manufacturing ecosystem, already primed for rapid innovation by two years of combat experience, moved with remarkable speed. The first Ukrainian fiber-optic drone flew less than three weeks after the first Russian fiber-optic strikes. The distributed nature of Ukrainian production, often criticized as inefficient, proved ideal for this kind of rapid adaptation. Dozens of manufacturers independently began experimenting with fiber-optic systems, each bringing slightly different approaches and innovations.

Artem Dzhulai, an engineer with Wild Hornets, a Kyiv-based drone manufacturer, led his company's fiber-optic program. "We didn't try to reverse-engineer the Russian system," he explained. "We looked at the basic principle, payload out cable, send light signals and designed from scratch. Within a week we had a prototype. Within two weeks we had something combat-ready. By week three, we were shipping production units to the front."

Ukrainian implementations quickly surpassed Russian designs in several key areas. While Russian systems used relatively heavy mechanical spools that paid out cable through friction-based mechanisms prone to jamming, Ukrainian engineers developed lighter weight spools with more reliable payout systems. Some used spring-loaded mechanisms that maintained consistent tension on the cable. Others employed weighted payouts where the cable's own weight provided smooth release.

The cable itself became an area of intense innovation. Early Ukrainian fiber-optic drones used standard telecommunications fiber, but this proved too fragile for combat conditions. The cable would break when the drone made aggressive maneuvers or when it snagged on vegetation, power lines, or battlefield debris. Manufacturers began sourcing reinforced fiber-optic cable with aramid fiber strengthening or even thin metal wire cores that could withstand substantial abuse while maintaining optical transmission.

Optical transceivers, the devices that converted electrical signals to light pulses and vice versa posed another engineering challenge. Off-the-shelf telecommunications transceivers were too heavy, power-hungry, and fragile for drone applications. Ukrainian engineers adapted components from fiber-optic communications equipment, miniaturizing and ruggedizing them for the harsh environment of drone warfare. Some clever manufacturers repurposed components from fiber-optic inspection equipment and medical endoscopes, finding existing technologies that could be adapted rather than designing from scratch.

The video transmission proved particularly tricky. Radio frequency video systems were mature technology with decades of development behind them. Sending video over fiber-optic cable required encoding the analog video signal digitally, transmitting it as light pulses, and decoding it at the ground station, all with minimal latency. Delays of even 100 milliseconds could make the drone unflyable at high speeds. Ukrainian engineers achieved latencies under 30 milliseconds through clever compression algorithms and high-bandwidth optical systems, making fiber-optic control feel as responsive as traditional radio control.

The Technical Deep Dive

Understanding the fiber-optic drone system requires examining each component and how they integrated into a functional weapons system. The technology represented a fascinating convergence of telecommunications, aerospace engineering, and combat adaptation.

The core of the system was the fiber-optic cable itself. Standard single-mode telecommunications fiber has a core diameter of about 9 micrometers, less than a tenth the width of a human hair. This tiny glass filament, protected by layers of cladding and jacketing, could transmit data at gigabit speeds over distances of many kilometers with virtually no signal loss. For drone applications, manufacturers used fiber with reinforced jacketing that could withstand the mechanical stresses of payout while maintaining optical clarity.

The spool mechanism demanded precision engineering. It needed to release cable smoothly as the drone flew away from the operator, maintaining consistent tension without creating drag that would affect flight performance. Too much tension and the cable would break. Too little and it would tangle or create excess slack that could foul the drone's propellers. The spool also needed to be light, every gram of weight reduced payload capacity or flight time.

Ukrainian manufacturers developed several spool designs. The most common used a passive friction system where the cable unwound against adjustable resistance, similar to a fishing reel. More sophisticated systems used active motorized spools that paid out cable at precisely controlled rates, sensing the drone's speed and adjusting payout accordingly. These active systems were heavier but prevented cable breakage and tangling more reliably.

The optical transceivers converted electrical signals from the flight controller and camera into modulated light that traveled down the fiber, and converted incoming light back into electrical signals the flight controller could use. These devices operated at wavelengths in the near-infrared spectrum, typically 850 nanometers or 1310 nanometers, optimized for transmission through multimode fiber over short distances. Data rates of 100 megabits per second or higher allowed full-resolution video and multiple control channels simultaneously.

Power represented a unique challenge. The drone required electrical power for its motors, flight controller, camera, and optical transceiver. But the fiber-optic cable carried only light, no electricity. This meant the drone had to carry all its power onboard in batteries, just like a radio-controlled drone. However, the ground station also required power for its optical transceiver and video display. This actually simplified ground station design, operators no longer needed powerful radio transmitters that consumed significant power and generated detectable emissions.

The flight controller firmware required modifications to work with fiber-optic control. Radio-controlled drones used bidirectional communication, the pilot sent commands and the drone sent back telemetry data about its position, battery level, and system status. Fiber-optic cables easily supported bidirectional communication by using two fibers (one in each direction) or by multiplexing bidirectional data on a single fiber using different wavelengths or time-division multiplexing.

Ukrainian engineers implemented telemetry systems that gave pilots the same information as radio systems, maintaining the situational awareness pilots depended on.

Tactical Advantages and Limitations

The fiber-optic solution wasn't perfect, but its advantages were compelling enough that both sides rapidly adopted it for specific mission profiles. Understanding when to use fiber-optic versus traditional radio-controlled drones became an essential part of tactical planning.

The most obvious advantage was complete immunity to radio frequency jamming. In heavily contested electromagnetic environments where traditional drones simply couldn't operate, fiber-optic drones flew with impunity. This made them ideal for attacking electronic warfare systems themselves, the very installations that would jam radio-controlled drones were helpless against fiber-optic attacks. Ukrainian forces used this capability to systematically degrade Russian electronic warfare networks, destroying jamming equipment with fiber-optic drones and then following up with conventional drones once the jamming lifted.

Fiber-optic drones also offered complete communications security. Radio transmissions, even encrypted ones, could be intercepted, direction-found, and analyzed. The physical cable meant there were no radio emissions to detect during flight. Ground stations became much harder to locate through electronic warfare means. Pilots could operate from positions that would be suicidal with radio-controlled drones, where their transmissions would draw immediate counter-battery fire.

The video quality over fiber-optic links surpassed radio systems. Without the interference, multipath fading, and bandwidth limitations of RF transmission, operators enjoyed crystal-clear high-definition video throughout the entire mission. This improved target identification and strike accuracy, reducing the chance of engaging friendly forces or civilians by mistake.

But fiber-optic guidance came with significant trade-offs that limited its universal adoption. The most fundamental was range. The cable could only be so long before its weight became prohibitive. Most Ukrainian fiber-optic drones carried 10-15 kilometers of cable, with some specialized long-range variants extending to 20 kilometers. Beyond that, the spool became too heavy and the drag too great. Radio-controlled drones could operate at much greater ranges, limited only by battery capacity and radio horizon.

The cable itself created tactical vulnerabilities. It left a visible trail back to the operator's position, literally a line leading straight to the pilot. Observers could follow the cable back to its source, potentially calling artillery or airstrikes on the ground station. The cable could also catch on obstacles: power lines, trees, buildings, even battlefield debris. A snagged cable could jerk the drone out of the sky or break entirely, terminating the mission.

Maneuverability suffered compared to radio-controlled drones. The cable created drag that increased with speed and the length of cable paid out. Aggressive turns could generate forces that broke the cable. The weight of the spool and cable reduced the payload capacity available for warheads or sensors. Pilots had to fly more conservatively, making fiber-optic drones less ideal for engaging small or moving targets that required aggressive pursuit.

The fiber-optic cable was also single-use. Once deployed, it couldn't be recovered, kilometers of fiber-optic cable trailed across the battlefield after each mission. This represented a recurring cost that traditional radio-controlled drones didn't incur. At scale, with thousands of daily drone missions, the cost of fiber-optic cable added up. Manufacturing couldn't always keep pace with consumption, creating supply bottlenecks.

Weather conditions affected fiber-optic drones differently than radio-controlled variants. High winds could catch the trailing cable, creating drag and control problems. Rain didn't affect the optical signals, but could make the cable heavier and more likely to catch on vegetation. Snow could cause the cable to accumulate ice, adding weight and creating resistance during payout.

Mission Profiles and Tactical Integration

By mid-2024, both Ukrainian and Russian forces had developed doctrine for when to use fiber-optic versus radio-controlled drones. The technologies weren't replacements but complements, each optimized for different tactical situations.

Fiber-optic drones excelled at attacks in heavily jammed environments. When assaulting well-defended positions ringed with electronic warfare systems, Ukrainian commanders would lead with fiber-optic drones to suppress or destroy the jamming equipment, then follow with waves of cheaper radio-controlled drones once the electromagnetic environment was permissive. This sequential approach maximized the cost-effectiveness of both technologies.

Counter-battery missions leveraged fiber-optic drones' resistance to jamming and lack of detectable emissions. Russian artillery and multiple rocket launcher positions typically deployed with electronic warfare protection and sophisticated counter-drone systems. Fiber-optic drones could penetrate these defenses, strike the artillery systems themselves, and withdraw without the ground station being located through radio direction finding.

Urban warfare became a primary mission set for fiber-optic drones. Cities' dense electromagnetic environments, with countless radio sources and reflective surfaces that created multipath interference, made radio control difficult. But fiber-optic cables could snake through urban terrain without signal degradation. Ukrainian forces used fiber-optic drones extensively in the defense of cities like Bakhmut and later in the Kursk offensive, where urban combat predominated.

Precision strikes against high-value targets justified the higher cost and operational complexity of fiber-optic systems. Assassinating senior officers, destroying command vehicles, or attacking key logistics nodes warranted using the more expensive but more reliable fiber-optic drones. Their immunity to jamming and superior video quality ensured mission success rates approaching 90%, compared to perhaps 30-40% for radio-controlled drones in contested environments.

Conversely, radio-controlled drones remained superior for certain missions. Long-range strikes beyond 15-20 kilometers required radio control. Attacks on moving targets benefited from radio-controlled drones' superior agility. Reconnaissance missions that required loitering over an area worked better with radio drones that didn't have a cable to manage. And simple mass attacks, where volume mattered more than individual success rates, favored cheaper radio-controlled drones.

Some units experimented with hybrid tactics, using radio-controlled drones for initial reconnaissance and target identification, then calling in fiber-optic strikes once targets were located. This preserved the more expensive fiber-optic drones for high-confidence engagements while using radio drones for riskier search and surveillance missions.

The Innovation Race Continues

Neither side rested after initial fiber-optic deployments. The technology sparked another round of innovation as engineers sought to overcome its limitations while preserving its advantages.

Russian developers began experimenting with breakaway systems that allowed the drone to detach from the cable during the final attack phase. The idea was to trail the cable during the vulnerable approach through jamming zones, then cut loose and complete the attack as a free-flying munition. This theoretically combined fiber-optic immunity during approach with radio control's agility during final engagement. In practice, the systems proved unreliable, the transition from fiber to radio control took precious seconds during which the target could evade or air defenses could engage.

Ukrainian engineers pursued increased cable length through lighter-weight fibers and more efficient spool mechanisms. By summer 2024, experimental systems achieved 30-kilometer cable runs, though these required larger drone platforms with greater lifting capacity. The longest-range fiber-optic drones began approaching the capabilities of loitering munitions, capable of striking strategic targets deep behind front lines with complete jamming immunity.

Automated cable management systems represented another innovation frontier. These systems used sensors to detect cable tension, payout rate, and potential snags, adjusting the spool release in real-time to prevent breakage. Some incorporated artificial intelligence algorithms that learned optimal payout profiles for different flight patterns and speeds. The automation reduced pilot workload and improved mission success rates, though at the cost of additional weight and complexity.

Cable recovery became a focus for environmental and operational security reasons. Kilometers of fiber-optic cable left across the battlefield could reveal flight paths and operational patterns to enemy intelligence.

Biodegradable cable coatings were developed that would cause the fiber to deteriorate within days of exposure, though these reduced tensile strength. Some manufacturers experimented with extremely thin fibers that would be nearly invisible even when fresh.

Hybrid fiber-radio systems emerged that used fiber-optic cable as primary control but incorporated backup radio systems. If the cable broke during a mission, the drone would automatically switch to radio control and attempt to complete its mission or return to base. These systems provided redundancy but at the cost of reintroducing radio detectability and vulnerability to jamming.

The Strategic Impact

The fiber-optic solution's significance extended beyond tactical advantages. It represented a broader lesson about technological adaptation in modern warfare. When electronic warfare threatened to make drone warfare obsolete, engineers didn't develop more sophisticated electronics, they bypassed electronics entirely by reaching back to a centuries-old principle: physical connections can't be jammed.

This pattern, solving modern problems with elegantly simple solutions rather than added complexity, recurred throughout the drone war. It spoke to the adaptive advantage of distributed, entrepreneurial defense industries over centralized bureaucratic systems. A dozen Ukrainian workshops could try a dozen different approaches to fiber-optic guidance, rapidly identifying what worked through battlefield testing. Traditional procurement systems would still be writing requirements documents while combat units were already flying fifth-generation fiber-optic drones.

The technology's rapid proliferation also demonstrated the impossibility of maintaining technological advantages in modern conflicts. Russia deployed fiber-optic drones in March. Ukraine had working copies in April. By June, both sides were flying second-generation systems with improved capabilities. The innovation cycle had compressed to weeks, making secrecy nearly impossible and technological leads ephemeral.

For military planners worldwide, the fiber-optic solution provided lessons about the future of warfare in contested electromagnetic environments. Space-based and electronic systems that militaries had relied on for decades could be jammed, spoofed, or destroyed. Physical connections, whether fiber-optic cables, hardened landlines, or other analog systems might prove more reliable in high-intensity conflicts. The future might look surprisingly like the past, with modern technology grafted onto old principles.

Living with the Wire

By late 2024, fiber-optic drones had become routine across the Ukrainian front. Units maintained mixed fleets of fiber-optic and radio-controlled drones, selecting the appropriate tool for each mission. Pilots trained on both systems, learning the different handling characteristics and tactical considerations of each.

The battlefield changed in subtle ways. Abandoned fiber-optic cables littered combat zones, thin glass threads stretching across no-man's-land like the web of some demented spider. Engineers picking through destroyed Russian positions found spools of fiber-optic cable alongside ammunition and supplies, testament to how central the technology had become. Both sides' electronic warfare operators learned to differentiate the radar signatures of fiber-optic versus radio-controlled drones, though they remained equally helpless against the former.

Dmitry Koval, the engineer who had recognized the elegance of fiber-optic guidance when Russian drones first appeared, reflected on how quickly the revolutionary became routine. "Six months ago, we were celebrating each successful fiber-optic mission. Now it's just another tool in the kit. Pilots barely think about whether they're flying fiber or radio, they just grab whichever drone suits the mission. That's how you know a technology has truly arrived: when it becomes boring."

The fiber-optic solution hadn't ended the electronic warfare arms race. Both sides continued developing more sophisticated jamming systems, better encryption, and more resilient radio protocols. But fiber-optic drones ensured that even the most intense electronic warfare couldn't shut down drone operations entirely. The sky over Ukraine would remain full of drones, trailing their thin glass lifelines back to operators who could now fight through any jamming the enemy could generate.

In a war defined by rapid technological adaptation, the fiber-optic breakthrough stood out not for its complexity but for its elegant simplicity. A technology developed for peaceful telecommunications, adapted to guide weapons of war, had become indispensable in a matter of months. The wire trailing behind each drone represented both a physical tether and a metaphorical link, connecting combat innovation to civilian technology, tactical needs to engineering solutions, and the present to an uncertain future where warfare would never be quite the same.

Chapter 11

Rubicon and the Elite Units

The name appeared first in intercepted Russian communications in May 2024, "Rubicon." Ukrainian signals intelligence operators noted the designation with interest but little alarm. New unit names emerged constantly in the fluid organizational structures of both armies. What made Rubicon different became apparent only when Ukrainian forces began dying in ways that suggested something had fundamentally changed in Russian drone operations.

"They weren't just better," explained Captain Andriy Kovalenko, a Ukrainian mechanized infantry company commander who faced Rubicon during the Kursk offensive in August 2024. "It was like fighting a completely different enemy. The drones came in coordinated swarms. They anticipated our movements. They struck weak points we didn't know we had. My men had been fighting Russian drones for two years, and they were scared in a way I'd never seen before. These weren't amateur hour volunteers flying quadcopters. These were professionals."

Rubicon represented Russia's acknowledgment that drone warfare had evolved beyond something that could be conducted by hastily trained volunteers or conventional forces given new equipment. It required specialization, professionalization, and institutional knowledge. The unit's creation and Ukraine's response through its own elite drone formations marked the maturation of drone warfare from improvised innovation to established military discipline.

Genesis of an Elite Unit

The Russian military's path to creating Rubicon was tortuous, constrained by bureaucratic inertia, institutional resistance, and the inherent conservatism of military organizations facing technological disruption. Through 2022 and early 2023, Russian drone operations remained ad hoc affairs, with conventional units assigned drone operators who learned on the job, volunteers from civilian drone racing communities pressed into service, and captured or purchased commercial drones distributed without systematic training or doctrine.

This decentralized approach produced individual successes but systemic inefficiency. Some units developed effective drone tactics through battlefield experimentation. Others wasted resources on ineffective employment or lost drones to operator error and poor tactical planning. There was no mechanism to capture lessons learned, disseminate best practices, or identify and cultivate the most talented operators. Drone warfare expertise was scattered across hundreds of units with no institutional structure to concentrate and multiply combat power.

The military theory supporting elite drone units was straightforward: drone operations, like sniper operations, artillery targeting, or special forces missions, benefited from specialization. A dedicated drone operator who flew daily, studied enemy countermeasures, and refined tactics would be vastly more effective than a conventional soldier who operated drones as one of many duties. Concentrating the most skilled operators in dedicated units would create disproportionate combat effects while establishing centers of excellence that could train and equip the broader force.

Rubicon's formation in early 2024 drew from multiple sources. The core cadre came from existing Russian drone units that had distinguished themselves in combat. These veterans brought operational experience and tactical knowledge earned through months of combat against increasingly sophisticated Ukrainian defenses. To this foundation, Russian military intelligence added drone racing champions, civilian UAV pilots with extensive flight hours, and even video game players with demonstrated skills in situational awareness and rapid decision-making.

The unit's name Rubicon, the river Julius Caesar crossed when committing Rome to civil war suggested both the irrevocability of the decision to create professional drone units and perhaps an awareness that this represented a threshold moment in military affairs. Once you crossed into the era of specialized drone warfare, there was no going back to conventional operations.

Training and Selection

Rubicon's selection process was deliberately rigorous, designed to identify operators with the rare combination of technical skill, tactical judgment, and psychological resilience that elite drone warfare demanded. Candidates underwent weeks of assessment that combined flight skill testing, psychological evaluation, and tactical problem-solving exercises.

The flight skill assessments went far beyond basic proficiency. Candidates flew increasingly complex missions through electronic warfare environments, navigating jamming that would cripple ordinary operators. They conducted precision strikes against moving targets, often required to hit specific weak points on armored vehicles rather than simply impacting anywhere on target. They flew through simulated anti-drone fire, dodging obstacles and employing evasive maneuvers while maintaining course to target.

Tactical judgment separated good pilots from great ones. Rubicon selection exercises presented candidates with complex operational scenarios requiring rapid decision-making under pressure. Should they strike an ammunition truck or continue searching for the command vehicle? When facing multiple targets of equal value, which offered the optimal strike angle and highest probability of kill? When should they abort a mission to preserve the drone for a better opportunity? These decisions, made in seconds during actual combat, could mean the difference between decisive tactical victories and wasted resources.

The psychological dimension proved equally critical. Drone warfare's remote nature created unique stresses. Operators watched their targets die in high-definition video, close enough to see faces, read expressions, and observe the human reality of the destruction they delivered. Unlike artillery crews or bomber pilots operating at distance, drone operators experienced an uncomfortable intimacy with killing that produced high rates of psychological strain. Rubicon sought operators who could maintain effectiveness despite this proximity, who could kill repeatedly without either becoming reckless or freezing under the moral weight.

Training for selected operators was intensive and ongoing. Rubicon established training facilities equipped with simulators that replicated the electromagnetic environments of Ukrainian-controlled territory, allowing operators to practice against realistic jamming and electronic warfare conditions. They flew hundreds of hours against mock targets, developing muscle memory and situational awareness that would prove crucial in combat where hesitation meant death.

The unit also received priority access to the newest drone technology. While conventional Russian units flew increasingly obsolete platforms or struggled with supply shortages, Rubicon operated the latest fiber-optic drones, enhanced FPV platforms with extended range, and experimental systems incorporating artificial intelligence and autonomous capabilities. This technological edge, combined with superior training and operator skill, made Rubicon units formidable opponents.

Organizational structure reflected lessons learned from special operations forces. Rubicon operated in small teams, typically four to six operators supported by technical specialists, intelligence analysts, and security personnel. These teams were mobile and self-sufficient, capable of deploying rapidly to critical sectors where their expertise could prove decisive. The flat organizational hierarchy empowered junior operators to make tactical decisions without navigating layers of approval, enabling the rapid responses that drone warfare demanded.

Doctrine and Tactics

Rubicon's tactical doctrine represented a synthesis of two years of Russian combat experience, distilled and refined by the unit's most capable practitioners. Where conventional Russian drone operations often consisted of individual strikes launched opportunistically, Rubicon employed sophisticated coordinated tactics that multiplied their effectiveness.

The unit pioneered multi-drone strikes that overwhelmed Ukrainian air defenses through saturation. Rather than sending single drones against targets protected by electronic warfare and small arms fire, Rubicon would launch four, six, or even eight drones simultaneously from different directions. Ukrainian defenders faced impossible choices, engage one drone and the others would strike home or try to divide fire and risk hitting none. The tactics borrowed from anti-ship missile saturation attacks, adapted to the small-unit scale of drone warfare.

Rubicon teams developed hunter-killer tactics pairing reconnaissance and strike platforms. A reconnaissance drone, often a commercial quadcopter modified for extended endurance, would orbit over Ukrainian positions identifying targets. Strike drones, typically FPV kamikaze platforms with heavy warheads, would wait concealed kilometers away. When the reconnaissance platform identified a high-value target, strike drones would launch, guided by real-time video to precision impacts. This division of labor allowed extended surveillance without exposing expensive strike platforms to air defense unnecessarily.

The unit excelled at exploiting brief windows of Ukrainian vulnerability. Rubicon operators studied Ukrainian unit routines, identifying patterns in shift changes, supply runs, and command movements. They noted when Ukrainian electronic warfare systems cycled for maintenance or battery changes, creating gaps in coverage. These observations informed precisely timed strikes that caught Ukrainians during their most vulnerable moments, during resupply, during leadership briefings and during the critical minutes when defensive systems were offline.

Rubicon also developed counter-drone tactics for engaging Ukrainian UAVs. Using modified FPV drones equipped with shotgun-style flechette warheads or small explosive charges, Rubicon operators hunted Ukrainian reconnaissance drones. Destroying these surveillance platforms blinded Ukrainian artillery and prevented targeting of Russian positions. The counter-drone missions required extraordinary pilot skill, engaging small, maneuverable targets moving at high speeds but paid tactical dividends by allowing Russian forces to move and operate without constant aerial observation.

Perhaps most significantly, Rubicon pioneered psychological warfare through drone operations. They deliberately targeted Ukrainian officers, unit leaders, and key specialists, seeking to decapitate command structures and demoralize survivors. Videos of successful strikes against Ukrainian leadership were disseminated through Russian military channels, intended to create fear and undermine morale. The tactic was ruthless but effective, Ukrainian officers learned to avoid predictable patterns, limit time in exposed positions, and became more cautious in ways that degraded operational tempo.

Kursk: Rubicon's Proving Ground

The Kursk offensive in August 2024, Ukraine's bold cross-border thrust into Russian territory provided Rubicon its first large-scale test. Ukrainian forces, having achieved tactical surprise, pushed rapidly into Kursk Oblast, capturing territory and drawing Russian forces into defensive operations on their own soil. The offensive represented Ukraine's most significant military success in months, and the Russian response would demonstrate how elite drone units could influence conventional operations.

Rubicon deployed to Kursk within 48 hours of the offensive's launch, establishing operating positions near the fluid front lines. Their mission was to degrade Ukrainian command and control, interdict supply lines, and provide reconnaissance that would enable Russian counterattacks. The unit brought extensive technical advantages, newer drones, better training, superior tactics but faced Ukrainian forces who had their own two years of drone warfare experience.

The impact was immediate and devastating. Ukrainian mechanized columns advancing along highways found themselves under relentless drone attack from Rubicon teams who had positioned reconnaissance platforms overhead. The Russians identified command vehicles, attacked supply trucks carrying ammunition and fuel, and struck Ukrainian combat engineering equipment needed to breach defensive positions. Where Ukrainian forces expected the sporadic drone harassment they'd grown accustomed to, they instead faced sustained, professional drone operations that systematically dismantled their offensive capabilities.

Captain Kovalenko's mechanized company experienced Rubicon's attention during the third week of the offensive. "We'd taken a small village and were consolidating, preparing to push forward," he recalled. "Then the drones came. First one, which we shot down. Then three more from different directions. We got two, but the third hit our BTR-4 and killed the crew. Before we could react, another wave came in low, using buildings for cover. One struck our ammunition truck. The explosion killed six men and left us critically short on supplies."

What particularly unnerved Kovalenko and his men was Rubicon's apparent omniscience. "They knew where to hit us. Not just random attacks, they went after our command vehicle, our logistics, our heavy weapons. It felt like they were inside our planning sessions. Later, we realized they'd been watching us for hours with reconnaissance drones we never spotted, mapping out our positions, identifying key targets. By the time the strike drones arrived, they knew exactly what to hit."

Ukrainian casualties mounted. Accurate statistics were difficult to obtain during the chaotic offensive, but battalion commanders reported that drone strikes, particularly those exhibiting Rubicon's characteristics caused 30-40% of Ukrainian losses in some sectors. More significantly, the constant drone threat slowed Ukrainian operations. Forces that should have been advancing aggressively instead moved cautiously, dispersed more widely, and spent precious time on air defense that could have been devoted to offensive action.

Rubicon's reconnaissance capabilities proved equally significant. Their continuous surveillance of Ukrainian positions provided Russian commanders unprecedented situational awareness. Russian artillery strikes arrived with uncanny accuracy, often catching Ukrainian forces during brief moments of exposure. Russian counterattacks struck Ukrainian weak points that conventional reconnaissance might never have identified. The information advantage that elite drone units provided was translating directly into combat power.

Ukrainian Response

Ukraine's military leadership, observing Rubicon's effectiveness, accelerated their own programs for creating elite drone units. Ukrainian forces had always maintained a degree of specialization in drone operations, certain battalions and brigades had developed reputations for drone expertise but nothing approaching Russia's formalized approach with Rubicon.

Several elite Ukrainian drone units emerged or received enhanced status during and after Kursk. The "Birds of Magyar" unit, associated with the controversial commander Artem Shtogrin (callsign Magyar), had operated as an elite drone formation since 2023, gaining notoriety for precision strikes and innovative tactics. The "Wild Hornets" expanded beyond their manufacturing role to establish dedicated operator teams. New units with exotic names, "Night Witches," "Iron Owls," "Shadow Hunters" appeared in Ukrainian order of battle, each representing concentrations of elite drone operators.

These Ukrainian elite units drew operators from similar sources as Rubicon: combat veterans with proven drone skills, civilian racing pilots, and technically talented volunteers. Training programs, often established by the units themselves rather than through formal military channels, emphasized the same combination of flight skill, tactical judgment, and psychological resilience. Ukrainian elite drone operators developed their own sophisticated tactics, often inspired by observing Russian innovations but adapted to Ukrainian operational needs.

The Ukrainian approach remained characteristically decentralized. Where Russia concentrated expertise in specific units like Rubicon, Ukraine maintained multiple competing elite formations, each developing its own tactics and specialties. This created inefficiency, resources were divided rather than concentrated but also fostered innovation as different units experimented with different approaches. Some Ukrainian elite units specialized in deep-strike operations against rear-area targets. Others focused on defensive counter-drone operations. Still others developed expertise in urban warfare drone tactics.

Ukrainian elite units also benefited from closer integration with volunteer networks and civilian manufacturers. While Russian units operated within formal military supply chains, Ukrainian elite operators often maintained direct relationships with drone manufacturers, providing feedback that shaped new designs and receiving early access to innovations. This symbiosis between elite operators and elite manufacturers created rapid innovation cycles that kept Ukrainian drone capabilities competitive despite Russia's larger resources.

The cultural differences between Russian and Ukrainian elite drone units reflected broader differences in military organization. Rubicon operated within Russia's hierarchical military structure, with clear chains of command, formalized procedures, and centralized control. Ukrainian elite units often maintained more anarchic structures, with operators enjoying greater autonomy and leadership emerging through demonstrated competence rather than formal rank. Both approaches had advantages, Russian centralization enabled coordinated large-scale operations, while Ukrainian decentralization fostered initiative and rapid adaptation.

The Evolution of Drone Tactics

The emergence of elite units on both sides accelerated the evolution of drone tactics, creating an innovation race where each side's advances forced rapid counter-innovations. The competition played out across multiple domains simultaneously.

Electronic warfare tactics became increasingly sophisticated. Elite operators learned to identify and avoid enemy jamming zones, to exploit brief gaps in coverage, and to employ techniques that minimized vulnerability to electronic attack. Some operators developed the ability to "read" the electromagnetic environment, sensing jamming intensity through subtle changes in control responsiveness and video quality, then adapting their approach accordingly. Others mastered rapid frequency changes, switching channels during flight to evade narrowband jamming.

Camouflage and concealment tactics evolved to counter drone reconnaissance. Elite units on both sides became adept at identifying the subtle signatures that revealed enemy positions, vehicle tracks, thermal signatures from equipment or personnel, electronic emissions from radios or jamming systems. This forced constant adaptation by their opponents. Units learned to minimize heat signatures, to use camouflage netting designed to defeat both visual and thermal detection, to employ decoys that drew drone strikes away from real assets.

Counter-drone tactics became a specialized domain within drone warfare. Elite units developed expertise in engaging enemy drones, using everything from small arms fire to purpose-built anti-drone drones. They learned to predict enemy drone approach routes, to identify the acoustic signatures of different drone types, to employ electronic warfare against enemy control signals. The most elite operators could identify threats from fragmentary information, a brief glimpse of a distant drone, a characteristic motor sound, a suspicious movement pattern on the horizon and respond before the threat materialized.

Coordination between drone units and conventional forces reached new levels of sophistication. Elite drone operators integrated into combined arms operations, providing reconnaissance, target designation, and direct fire support in coordination with infantry, armor, and artillery. They learned to communicate effectively with commanders who often lacked technical understanding of drone capabilities and limitations, translating tactical requirements into drone missions that achieved operational objectives.

The Human Factor

Behind the tactical evolution were human beings adapting to warfare's newest dimension. Elite drone operators occupied a unique psychological space, technical specialists who were also combat arms soldiers, remote warriors who nonetheless experienced intimate proximity to the killing they conducted. The psychological toll was substantial and ongoing.

Vasily, a Rubicon operator who agreed to speak on condition of anonymity, described the emotional complexity: "I've killed more men than I can count. I've watched them die on my screen, close enough to see their faces. Sometimes they see the drone coming, I watch them try to run or take cover, or just freeze in terror. Then the impact, and they're gone. At first it affected me deeply. I had nightmares, and couldn't sleep. Now it's just part of the job. I don't know if that's better or worse."

The intimacy of drone killing created unique moral challenges. Artillery crews might fire hundreds of rounds never seeing their targets. Pilots dropped bombs from altitudes where individual humans were invisible. But drone operators watched their targets in high definition, saw them as individuals, and delivered death with precision that eliminated any psychological distance. Some operators developed coping mechanisms, emotional detachment, viewing targets as game pieces rather than humans, dark humor shared with fellow operators. Others struggled with growing moral injury that manifested as depression, substance abuse, or operational degradation.

Elite units attempted to address psychological strain through rotation policies and peer support systems. Rubicon mandated regular psychological assessments and enforced rest periods where operators stepped away from combat flying. Ukrainian elite units, with their less formal structures, relied more on informal peer support, operators looking out for each other, identifying when comrades were struggling, and intervening before problems became incapacitating.

The operators also faced physical stresses that were easy to overlook. Hours spent hunched over controllers in confined spaces caused chronic back and neck problems. Eye strain from staring at small screens in varying light conditions became endemic. Poor sleep, irregular meals, and sustained stress took cumulative tolls on health. The warfare was remote but not without physical cost.

Yet many operators described their role with pride and sense of purpose. They were pioneers in a new form of warfare, writing doctrine through combat, developing capabilities that would reshape military operations for decades. "We're not just fighting this war," explained Dmytro, a Ukrainian elite drone operator. "We're inventing how future wars will be fought. That's historic. It's also terrifying. But there's pride in knowing that what we're learning, what we're doing, will matter long after this conflict ends."

Organizational Challenges

The professionalization of drone warfare through elite units created organizational challenges that both sides struggled to resolve. Where should these units fit in military hierarchies? Should they be independent formations or attached to conventional units? How should resources be allocated between elite drone units and conventional forces that also needed drones?

Russia's solution with Rubicon was to create a separate force structure with independent chains of command reporting directly to higher headquarters. This gave Rubicon operational flexibility and protected their resources from being diverted to conventional units. But it also created friction with conventional commanders who wanted elite drone support but lacked authority to task Rubicon units. Coordination problems emerged where Rubicon operations conflicted with conventional forces' plans, or where elite units identified targets but conventional forces lacked the responsiveness to exploit the intelligence.

Ukraine's decentralized approach created different problems. Multiple elite units competed for resources, attention, and missions. Coordination between elite drone units and conventional forces varied wildly, some brigades integrated elite drone teams seamlessly, while others viewed them as unreliable cowboys pursuing their own agendas. The lack of standardization meant tactics and capabilities differed significantly between units, making it difficult for commanders to know what to expect from elite drone support.

Both sides grappled with how to scale elite capabilities. Rubicon could only be so large before it lost the selectivity that made it elite. But the demand for elite-level drone operations far exceeded what a single unit could provide. Training additional elite units risked diluting quality as the pool of exceptional talent was finite. Some Russian military theorists advocated for a tiered system, a small core of elite units like Rubicon, surrounded by proficient conventional drone units trained in tactics developed by the elite formations, all supported by mass-production low-skill-threshold drones for saturation attacks.

Resource allocation created constant tension. Elite units consumed disproportionate shares of advanced drones, training resources, and support assets. Conventional commanders resented elite formations' privileged positions, arguing that distributing resources more evenly would provide better overall combat power. Elite unit advocates countered that concentration of expertise and equipment generated disproportionate effects that one elite operator with superior equipment was worth ten conventional operators with standard kit.

Legacy and Lessons

By late 2024, elite drone units had become permanent features of both Russian and Ukrainian force structures. Rubicon and its Ukrainian counterparts demonstrated that specialization in drone warfare generated significant combat advantages. The lesson extended beyond this specific conflict, future military forces worldwide would need to decide how to organize, train, and employ their own elite drone formations.

The emergence of units like Rubicon represented the maturation of drone warfare from improvised innovation to established military discipline. The anarchic early days, where anyone with a quadcopter could contribute to the war effort, were giving way to an era where professional specialists dominated the most critical missions. This professionalization was inevitable as with any military technology, initial improvisation eventually yields to doctrine, organization, and specialization.

Yet the tension between elite specialization and mass employment remained unresolved. Elite units like Rubicon demonstrated what highly trained operators with advanced equipment could achieve. But the broader war still consumed thousands of cheap drones daily, flown by conventional soldiers with basic training. Both approaches were necessary. Elite units conducted precision missions that conventional operations couldn't accomplish. Mass drone employment created the sustained pressure that shaped the battlefield.

Captain Kovalenko, reflecting on his unit's encounters with Rubicon during Kursk, captured both the fear and respect that elite drone units inspired: "Facing them was terrifying. But it also showed us what we needed to become. We couldn't match their resources or their training in the short term. But we could learn from their tactics, we could improve our own skills, we could become harder targets. The competition made us better. And that's war, constant adaptation or death. Rubicon adapted faster than most. That's why they were elite."

As 2024 progressed into 2025, elite drone units on both sides continued their deadly competition. Each innovation spawned counter-innovations, each tactical success prompted tactical adjustments, each victory in the electromagnetic battlespace created new challenges to overcome. Rubicon and its Ukrainian counterparts were locked in an endless dance of adaptation, writing the doctrine of drone warfare through combat, one mission at a time. The future of warfare was being forged in their hands, guided by their eyes, and delivered by their machines, elite warriors for an unmanned age.

Chapter 12

The Saturation Campaign

The air raid siren began at 11:47 PM, its wailing tone cutting through the September night in Kyiv. Iryna Bondarenko, a 34-year-old teacher, barely stirred in her bed. This was the third alert that night, the nineteenth that week. The rhythm had become grimly familiar: the siren's rise and fall, the automated voice on the emergency broadcast warning of incoming aerial threats, the decision whether to go to the shelter or stay in bed. Increasingly, exhausted Kyivans were choosing sleep over safety.

But this night was different. At 11:52 PM, the first explosion echoed across the city. Then another. And another. What began as routine became catastrophe as wave after wave of Russian Shahed drones descended on Ukraine's capital and cities across the nation. By dawn, when the all-clear finally sounded, Ukraine would record one of the heaviest single-night drone attacks of the war, over 120 drones launched against civilian infrastructure, power facilities, and population centers.

"I stopped counting the explosions after twenty," Iryna recalled. "I was in the shelter with my neighbors, and we could hear them. Some close, some distant, but constant. Someone kept saying 'it has to stop soon, they can't have that many.' But they did. They just kept coming. That's when I understood, this wasn't about destroying specific targets anymore. This was about breaking us."

The saturation campaign that Russia launched in mid-2024 represented a strategic shift in drone warfare. Where earlier strikes had been opportunistic or targeted at military objectives, the new campaign employed drones as instruments of terror, attacking the psychological and economic foundations of Ukrainian society through sustained, systematic nighttime bombardment. It was strategic bombing reimagined for the drone age, cheap, persistent, and designed not to win battles but to destroy the will to resist.

The Strategic Logic

The decision to launch sustained saturation drone attacks against Ukrainian cities emerged from Russia's strategic frustration. By mid-2024, the ground war had calcified into attritional stalemate. Russian forces had captured Bakhmut and made incremental gains in Donetsk Oblast, but at horrific cost. Ukrainian defenses, reinforced by Western aid and increasingly sophisticated in their employment of drones and electronic warfare, had prevented any breakthrough that might lead to operational or strategic success. Moscow needed a new approach.

Drones offered what conventional airpower couldn't. Russian piloted aviation, bloodied by Ukrainian air defenses in the war's opening months, had been relegated to launching standoff weapons from Russian airspace, rarely penetrating Ukrainian territory. But drones, particularly the Iranian-designed Shahed-136 that Russia was now producing domestically in substantial numbers could be launched by the hundreds at costs measured in tens of thousands of dollars per unit. Even with high attrition rates, enough would penetrate defenses to strike their targets.

The strategic calculus was brutally simple. Ukraine's air defense network, designed to counter aircraft and ballistic missiles, struggled against the saturation tactics that swarms of cheap drones enabled. A Patriot missile costing $4 million might shoot down a Shahed worth $20,000, an exchange ratio that heavily favored the attacker. Ukraine could shoot down 80% or even 90% of incoming drones and still suffer significant damage from those that penetrated. The mathematics of attrition favored Russia's deeper resources and larger production capacity.

More insidiously, the drone campaign targeted Ukrainian civilian morale and economic sustainability. Nightly attacks on power infrastructure created cascading blackouts that affected millions. Strikes on water treatment facilities threatened public health. Attacks on transportation networks complicated logistics. And the constant psychological pressure of nightly attacks, the interrupted sleep, the fear and the economic disruption all of it eroded the home front's capacity to sustain the war effort.

Russian strategic planners drew lessons from World War II strategic bombing campaigns, particularly the German V-1 and V-2 attacks on London. Those campaigns had failed to break British morale but had forced Britain to divert enormous resources to defense, reconstruction, and civil defense. The modern drone campaign sought similar effects: even if it didn't break Ukrainian will, it would impose costs that complicated Ukraine's war effort and potentially eroded Western support by creating a perception that the war was unwinnable.

The Campaign Begins

The saturation campaign escalated gradually through summer 2024 before reaching full intensity in autumn. Russian drone production, having achieved industrial scale by late 2023, now generated sufficient numbers to sustain nightly attacks. Iran had transferred production technology allowing Russia to manufacture Shahed variants domestically, insulating the supply from international sanctions and diplomatic pressure. By August 2024, Russian production was estimated at 300-400 drones per month, with projections suggesting capacity would reach 500 or more by year's end.

The attacks followed predictable patterns that Ukrainian defenders learned to anticipate but struggled to counter. Launches typically began in the late evening, with drones departing from bases in southern Russia, Belarus, and occupied Crimea. Flying at low altitudes and relatively slow speeds, around 180 kilometers per hour the drones approached Ukrainian cities in waves, often separated by 15-30 minutes. This spacing forced Ukrainian air defenses to maintain alert status throughout the night, exhausting operators and depleting ammunition stocks.

The drones themselves were crude but effective weapons. The Shahed-136 and its Russian-produced variant, the Geran-2, consisted of plywood and foam construction with a small gasoline engine providing propulsion. A warhead of approximately 50 kilograms of high explosive gave each drone destructive power equivalent to a 155mm artillery shell, enough to destroy buildings, ignite fires, and kill anyone unfortunate enough to be near the impact point. Simple GPS guidance took the drones to their target coordinates, though accuracy was poor by precision weapon standards, with circular error probabilities measured in tens of meters.

What the drones lacked in sophistication they compensated for through sheer numbers. A typical major attack involved 50-100 drones launched against multiple cities simultaneously. Kyiv, Kharkiv, Odesa, Dnipro, and other population centers faced coordinated strikes designed to overwhelm localized defenses. The attacks demonstrated strategic coordination, drones approached from multiple directions, forcing defenders to divide their attention and resources. Some drones carried radar reflectors or electronic warfare equipment, creating false targets that complicated defensive calculations.

Ukrainian air defenses achieved remarkable success rates given the challenge. Mobile air defense teams using everything from Soviet-era systems to modern Western equipment hunted the slow-moving drones, shooting them down with surface-to-air missiles, anti-aircraft artillery, and even machine gun fire. Fighter aircraft scrambled to intercept drones beyond the range of ground-based systems. On many nights, Ukraine claimed destruction of 80% or more of incoming drones, an impressive defensive performance by any standard.

But the 10-20% that penetrated defenses were enough. A single drone striking a power substation could plunge hundreds of thousands into darkness. One hitting a fuel depot created spectacular fires and deprived the economy of critical resources. A drone impacting a residential building killed civilians and generated the terror and outrage that Russia sought. The attackers needed only partial success to achieve their strategic objectives.

Life Under the Drones

For Ukrainian civilians, the saturation campaign transformed daily existence in ways large and small. The predictability of nighttime attacks meant that by 10 PM, cities began preparing for the inevitable sirens. Families charged phones and portable batteries, filled water containers in case supplies were disrupted, and prepared shelter spaces in basements or interior rooms. Some kept bags packed with essentials, documents, medications and warm clothing ready to grab when sirens sounded.

The shelters themselves became nocturnal communities. Kyiv's metro system, built during the Soviet era with bomb shelters in mind, served as a primary refuge for hundreds of thousands. Each night, Kyivans descended into the underground, spreading blankets on platforms, trying to sleep on hard concrete while waiting for the all-clear. Children did homework by flashlight. Elderly residents struggled with stairs and cold. Young couples held each other, listening to distant explosions, wondering if their apartments would survive the night.

Sleep deprivation became an epidemic. Even those who chose to stay in their apartments rather than shelter found sleep nearly impossible. The sirens alone were designed to wake the dead, piercing, impossible to ignore. The tension of waiting for impacts, the distant thunder of explosions or anti-aircraft fire, the uncertainty about whether the next blast would be closer, all of it made rest elusive. People went to work exhausted, their productivity degraded by accumulated fatigue. Accidents increased. Tempers shortened. The psychological toll mounted.

Economic impacts rippled through society. Businesses couldn't operate effectively when employees were exhausted and infrastructure unreliable. International companies reconsidered operations in Ukraine when their offices might lose power multiple times daily. Tourism, already devastated by war, became nearly nonexistent as foreign visitors avoided cities under nightly attack. The drone campaign was achieving economic warfare objectives even when drones missed their targets, the disruption and fear alone imposed enormous costs.

Children paid particular prices. Schools adapted schedules around expected attacks, starting classes later to allow students recovery time after nighttime disruptions. Some lessons moved to basements equipped as shelters, so education could continue even during raids. Psychologists noted increasing anxiety disorders, sleep problems, and behavioral issues among children who had spent formative years under aerial threat. An entire generation was growing up knowing no life without air raid sirens.

Yet resilience emerged alongside suffering. Communities organized mutual aid networks, with neighbors checking on elderly residents, sharing resources during power outages, and supporting each other through difficult nights. Dark humor became a coping mechanism, jokes about "Russian alarm clocks" and memes about Shahed drones circulated on Ukrainian social media. Cafes advertised "bomb shelter happy hours." Bakeries promoted "air raid specials" for customers emerging from overnight alerts. Life continued, adapting to circumstances that would have seemed unimaginable three years earlier.

Iryna Bondarenko, the Kyiv teacher, described the psychological adaptation: "At first, every siren was terrifying. My heart would race, my hands would shake. After months, it became... not routine exactly, but familiar. You learn to function with constant threat. You don't stop being afraid, but the fear becomes background noise instead of overwhelming panic. I don't know if that's strength or just exhaustion. Maybe both."

The Infrastructure War

While the psychological impact of saturation attacks was significant, the tangible damage to Ukrainian infrastructure presented the more immediate strategic challenge. Russian targeting prioritized electrical generation and distribution, water treatment and supply, heating systems critical for winter survival, and transportation networks that sustained the economy and military logistics.

Ukraine's electrical grid became the primary target. Substations, transformers, and generating facilities faced repeated attacks designed to create cascading failures across the interconnected system. A single drone striking a key substation might cause blackouts affecting millions of people across multiple regions. The grid's age, much of it dating from Soviet times made it particularly vulnerable to disruption and difficult to repair with limited resources during wartime conditions.

Ukrainian power company technicians became unsung heroes of the campaign, working around the clock to restore service after attacks. They operated under dangerous conditions, knowing that repaired facilities would likely be targeted again. Replacement transformers and critical components, often sourced from Western allies, were installed only to be destroyed weeks or months later. The work was Sisyphean, endless repair of damage that would be inevitably renewed.

Water infrastructure represented equally critical targets. Treatment plants required electricity to function; attacks on power supplies simultaneously disrupted water services. Direct strikes on pumping stations and reservoirs threatened public health and basic survival. Cities implemented emergency measures, distributing water from tanker trucks, drilling temporary wells, educating residents on water purification. What had been taken for granted became daily challenges requiring conscious planning and effort.

Heating infrastructure, as autumn turned to winter, became targets of particular concern. Ukraine's district heating systems, which provided warmth to millions in apartment blocks across the country, depended on power plants and distribution networks vulnerable to drone strikes. Russian planners understood that disrupting heat in winter wasn't merely inconvenient, it was potentially lethal, especially for elderly or vulnerable populations. The campaign aimed to force an impossible choice: freeze or flee.

Transportation infrastructure faced systematic attack. Railway stations, bridges, and logistics hubs suffered repeated strikes. While military logistics remained primary targets, the attacks disrupted civilian transportation essential for economic activity and daily life. Fuel depots burned after drone impacts, creating shortages that rippled through the economy. Port facilities in Odessa faced particular attention as Russia sought to strangle Ukraine's remaining export capacity.

The cumulative effect was degradation of Ukraine's economic and social infrastructure faster than it could be repaired or replaced. Western aid provided resources for reconstruction, but workers couldn't repair faster than drones could destroy. Each attack imposed costs measured not just in immediate damage but in diversion of resources, money, materials, and labor from productive activity to constant repair and mitigation. The strategic grinding was achieving what Russian ground forces couldn't wear down Ukrainian capacity to sustain resistance.

The Defense Challenge

Ukrainian air defense forces faced overwhelming challenges. The scale of attacks demanded constant readiness, exhausting crews and systems. The geographic spread required defense assets distributed across the country, preventing concentration of forces. And the cost disparity, expensive missiles against cheap drones created unsustainable attrition even in successful engagements.

Ukraine employed layered defenses combining multiple systems. Long-range surface-to-air missiles like Patriots and NASAMs engaged drones at distance when possible, though using such expensive systems against cheap targets was economically painful. Medium-range systems like Gepard anti-aircraft guns provided point defense for critical facilities. Mobile teams with man-portable air defense systems (MANPADS) hunted individual drones. Even small arms fire proved effective against the slow, low-flying Shaheds when they passed within range.

Fighter aircraft played crucial roles in air defense, with Ukrainian MiG-29s and Su-27s intercepting drones beyond the reach of ground-based systems. Pilots developed tactics for engaging multiple drones during single sorties, using cannon fire to conserve expensive missiles. Night interceptions were particularly challenging, requiring radar guidance to locate drones invisible against dark skies. The operations were dangerous, flying low-altitude intercept missions at night over terrain where both Ukrainian and Russian air defenses might engage anything airborne.

Electronic warfare provided another defensive layer. Jamming systems could disrupt GPS guidance, causing drones to lose their way or crash before reaching targets. The effectiveness varied, some drones incorporated GPS backup systems or inertial navigation that maintained course despite jamming. The Shaheds' simple construction made them resistant to some electronic warfare techniques that worked against more sophisticated systems.

Innovation in air defense became a constant necessity. Ukraine developed shoot-down apps that allowed civilians to report drone sightings, creating distributed surveillance networks that supplemented military sensors. Acoustic detection systems identified drones by their distinctive engine noise. Mobile defense teams positioned throughout cities provided overlapping coverage that limited gaps in defensive umbrellas. Every advantage mattered when facing nightly saturation attacks.

Yet even successful air defense extracted costs. Ammunition expenditure was enormous, hundreds of missiles and thousands of shells fired monthly. Each shoot-down required fuel, maintenance, and crew time. Systems wore out under constant use, requiring replacement or extensive overhaul. Air defense personnel suffered exhaustion from endless night shifts maintaining alert status. The defense was winning tactically, shooting down most incoming drones while losing strategically as the campaign ground down Ukrainian resources and morale.

International Dimensions

The saturation campaign created diplomatic complications for Russia while generating international support for Ukraine. Drone attacks on civilian infrastructure violated laws of war that theoretically constrained state behavior. International humanitarian organizations condemned the campaign as illegal and immoral. Western governments denounced Russia's tactics and accelerated military aid to Ukraine, specifically including air defense systems designed to counter the drone threat.

But Russia calculated correctly that international condemnation meant little without enforcement. The United Nations Security Council, paralyzed by Russian veto power, couldn't act. Sanctions were already severe; threatening additional economic penalties had diminishing returns. Western military intervention remained politically unthinkable despite the civilian suffering. Russia could conduct saturation bombing with relative impunity, constrained only by Ukrainian defensive capabilities and its own production capacity.

The campaign did create risks of escalation. Drones strayed into neighboring countries' airspace with increasing frequency. Polish, Romanian, and Moldovan airspace was violated multiple times by Shaheds that missed their targets or were thrown off course by defensive measures. NATO members scrambled fighters to intercept wayward drones, creating dangerous situations where miscalculation could trigger Article 5 and draw the alliance into direct conflict.

These incidents generated tensions within NATO. Eastern European members, facing direct threats from Russian drones, advocated aggressive responses including shoot-down of any Russian aircraft approaching alliance airspace. Western European nations counseled caution, wary of an escalation spiral that might trigger wider war. The debate reflected fundamental tensions about how to support Ukraine without becoming party to the conflict, a line that stray drones constantly threatened to cross.

The campaign also tested Western resolve. Russia hoped that sustained attacks on Ukrainian cities would generate pressure on Kyiv to negotiate from Western allies wary of supporting endless war. If Ukraine couldn't defend itself against cheap drones despite massive military aid, what was the point of continued support? The strategy bet on Western fatigue, on publics grown tired of bearing costs for a distant conflict, on politicians looking for exits from costly commitments.

Ukraine countered by emphasizing both the humanitarian catastrophe and the broader implications. If Russia succeeded through saturation drone attacks, it would establish a precedent encouraging other nations to employ similar tactics. The defense of Ukraine was the defense of a principle that nations couldn't achieve strategic objectives through sustained terror bombing of civilians. Western support needed to continue not just for Ukraine's sake but to prevent the normalization of this form of warfare.

Psychological Operations

Beyond the physical damage, Russia employed the saturation campaign as psychological warfare against both Ukrainian society and its government. The attacks carried messages as well as warheads, communicating Russia's capacity to inflict suffering, Ukraine's inability to fully protect its population, and the futility of continued resistance.

Russian information operations amplified the psychological impact. State media broadcast footage of successful strikes, emphasizing damage and Ukrainian casualties. Telegram channels affiliated with Russian security services posted videos of burning buildings and devastated infrastructure, often with commentary suggesting Ukraine surrender rather than endure continued punishment. The message was clear: resistance brought only suffering, while submission might bring relief.

The targeting carried psychological calculations. Strikes on power infrastructure during coldest weather maximized discomfort and fear. Attacks timed to disrupt major holidays or national celebrations sought to demoralize. Occasional strikes on residential areas served as reminders that nowhere was truly safe. The psychological architecture was sophisticated, employing insights from behavioral psychology to maximize fear and helplessness.

Ukrainian society proved remarkably resistant to psychological manipulation. National unity, forged through years of Russian aggression, provided resilience against attempts to break morale. The attacks often backfired, strengthening resolve rather than weakening it. Each Russian strike reminded Ukrainians why they fought, reinforcing narratives of Russian brutality and Ukrainian victimhood that sustained domestic and international support.

But resistance came at psychological cost. Mental health professionals reported increasing rates of anxiety, depression, and post-traumatic stress as the campaign continued. Substance abuse increased. Domestic violence rose. Suicide rates climbed. The population was enduring trauma that would take generations to heal. Ukrainian society might resist breaking, but it was bending under sustained pressure in ways that would leave permanent scars.

The Endurance Contest

By late 2024, the saturation campaign had settled into a grim routine. Russian production delivered sufficient drones to sustain nightly attacks. Ukrainian defenses destroyed most but not all incoming threats. Infrastructure suffered ongoing degradation. Civilians adapted to life under siege, finding reserves of endurance they hadn't known they possessed. Neither side was winning, Russia wasn't breaking Ukrainian will and Ukraine wasn't stopping the attacks but both continued because neither could afford to be first to admit the strategy's failure.

The campaign illustrated fundamental truths about modern warfare. Cheap precision weapons, mass-produced and employed without concern for losses, could impose strategic effects against even well-defended opponents. Air defense, no matter how sophisticated, couldn't achieve perfect protection against saturation tactics. And civilians, however resilient, bore terrible costs in wars where the distinction between front line and home front had collapsed.

Iryna Bondarenko, the Kyiv teacher who had lived through hundreds of air raid alerts, reflected on what the campaign had taken and what remained: "They've destroyed buildings, killed my neighbors, ruined sleep, made us afraid. They've done real damage. But they haven't made us give up. Every morning after an attack, I see people going to work, children going to school, life continuing. That's victory, even if it doesn't feel like it. We're still here. We're still fighting. The drones fall, and we rebuild. That's all they've achieved, proving we won't break."

The saturation campaign continued into 2025, an endless rain of drones against cities that refused to surrender. The strategic question remained unanswered: would Ukraine's capacity to endure outlast Russia's capacity to produce and launch? The answer would shape not just this war but the future of conflict in an age where terror from the sky came not from billion-dollar bombers but from drones built in factories and launched by the hundreds into the darkness.

Chapter 13

Numbers That Matter

In the sterile conference room of Ukraine's Ministry of Defense, a large digital display showed numbers that told the story of modern warfare better than any tactical map or strategic assessment. The figures updated in near real-time, fed by air defense commands, intelligence units, and frontline reports. By the end of September 2024, the display showed a statistic that captured both the scale and the absurdity of the drone war: 5,636 Russian drones launched against Ukraine since the beginning of the year. Next to it, another number: 187 missiles. The ratio of 30 drones for every missile revealed how profoundly warfare had changed.

"When I first joined military intelligence in 2015," explained Colonel Mykola Danilov, a senior analyst with Ukraine's Main Intelligence Directorate, "we tracked missiles. Cruise missiles, ballistic missiles, air-to-surface missiles. Those were the threats that mattered. Drones were curiosities, reconnaissance platforms, maybe occasionally armed with small munitions. Now look at this." He gestured at the display. "Five thousand drones in nine months. That's the war. That's what we're fighting."

The numbers that emerged from 2024's drone war told stories that narrative alone couldn't capture. They quantified industrial capacity, revealed strategic priorities, demonstrated the mathematics of attrition warfare, and illustrated the unprecedented scale of unmanned combat. Understanding these numbers meant understanding the conflict itself, not just the tactical engagements but the industrial, economic, and strategic dimensions that determined success or failure.

The Production Race

At the heart of the drone war lay production capacity, the ability to manufacture unmanned systems fast enough to replace combat losses and expand deployed numbers. By October 2024, both Ukraine and Russia had achieved industrial scale that would have seemed fantastical in 2022.

Ukraine's production figures told a story of remarkable transformation. From essentially zero domestic drone manufacturing in early 2022, Ukraine produced an estimated 1.5 million FPV drones during 2024. The number was staggering, more drones in a single year than most nations' militaries operated across their entire history. Production peaked in summer months at over 150,000 units monthly, tapering slightly in autumn but maintaining rates that would have been considered impossible just eighteen months earlier.

The October 2024 announcement by President Zelensky that Ukraine would produce 1.6 million drones annually represented not aspiration but reality. Ukrainian manufacturers had already achieved this production rate; the announcement simply formalized what industry insiders knew. More significantly, Zelensky announced plans to expand capacity to 4 million drones annually by 2025, quadrupling an already unprecedented production level.

TAF Drones, the largest Ukrainian manufacturer, exemplified the scale achieved. The company's production reached 40,000 drones monthly by late 2024, a volume that generated revenues exceeding $1 billion annually. A single company was producing more drones each month than most national militaries possessed in total. The numbers weren't just statistics, they represented tens of thousands of workers, hundreds of facilities, entire supply chains devoted to manufacturing instruments of death by the millions.

Russian production figures, derived from intelligence assessments and battlefield evidence, suggested roughly comparable scale. Russia produced an estimated 1.3-1.5 million tactical FPV drones during 2024, slightly behind Ukrainian output but within the same order of magnitude. For Shahed-style loitering munitions, Russian production reached approximately 4,000-5,000 units during 2024, with monthly production rates increasing steadily through the year.

The parity in production rates between Ukraine and Russia represented a strategic achievement for Ukraine and a strategic failure for Russia. Ukraine, with a GDP roughly one-tenth Russia's size and a population less than one-third as large, had matched its adversary's drone production through distributed manufacturing, volunteer networks, and entrepreneurial initiative. Russia, with vast Soviet-era military-industrial infrastructure and far deeper resources, should have dominated production. That it didn't suggested fundamental problems in Russian defense industry organization and allocation of resources.

Both sides' production rates were sustained by massive financial commitments. Ukraine's $2.47 billion drone procurement contract for 1.6 million units represented approximately 3-4% of the nation's entire 2024 defense budget devoted to a single weapons category that hadn't existed in military arsenals five years earlier. Russia's spending on drone production remained opaque, but intelligence estimates suggested comparable or higher absolute expenditures given Shahed production costs and FPV drone manufacturing at scale.

The Consumption Rate

Production numbers mattered only in relation to consumption, how quickly drones were being destroyed, expended, or rendered inoperable. The mathematics of drone warfare were unforgiving: production had to exceed consumption or stockpiles would deplete, capabilities would degrade, and operational effectiveness would collapse.

Consumption rates proved difficult to quantify precisely, but battlefield evidence and unit reports provided reasonable estimates. Ukrainian forces were consuming approximately 10,000-15,000 FPV drones weekly during peak combat periods in summer and autumn 2024. This meant that Ukraine's monthly production of 150,000 drones barely exceeded or during intense fighting, slightly lagged monthly consumption. The balance was precarious, leaving little room for stockpiling or recovering from production disruptions.

Russian consumption rates appeared roughly similar based on Ukrainian observations of drone activity. Russian forces launched thousands of FPV drones weekly against Ukrainian positions, with attrition rates that varied by sector but averaged 70-80% mission loss rates when including drones destroyed by electronic warfare, shot down by air defenses, or lost to operator error and technical failures. Only 20-30% of launched drones achieved their intended effects, meaning that inflicting significant damage required launching many more drones than would ultimately strike targets.

The Shahed loitering munitions showed different consumption patterns. Russia launched approximately 5,636 Shaheds through September 2024, averaging roughly 625 per month. But monthly figures varied wildly based on operational tempo and strategic priorities. July 2024 saw the highest monthly total over 6,000 drones launched according to some Ukrainian assessments, though this figure likely included FPV drones and other platforms mixed with Shaheds. More conservative estimates suggested 500-600 Shaheds launched in July specifically.

The arithmetic was relentless and revealing. If Russia launched 500-600 Shaheds monthly and produced 400-450 monthly, stockpiles were declining. Either production had to increase, consumption had to decrease, or the campaign would eventually exhaust available drones. Russia appeared to be choosing increased production, with new manufacturing facilities coming online through late 2024 to expand capacity.

Ukraine's consumption of air defense munitions presented equally stark mathematics. Each successful shoot-down of a Russian drone required missiles, shells, or bullets resources that were finite and expensive. Ukraine fired thousands of air defense missiles during 2024, depleting stockpiles faster than Western allies could replenish them. The cost asymmetry was brutal: a $20,000 Shahed drone forced the expenditure of missiles costing $50,000 to $4 million depending on the system. Russia could sustain economic attrition rates that bankrupted Ukrainian air defense.

The Kill Ratios

Numbers quantified not just production and consumption but effectiveness, how many drones achieved their intended effects versus how many were neutralized before impact. These kill ratios revealed the effectiveness of defenses and informed strategic calculations about whether drone campaigns achieved worthwhile results.

Ukrainian air defense achieved remarkable success rates against Russian drone attacks. Throughout 2024, Ukraine consistently reported destroying 70-80% of incoming Shahed drones, with some particularly effective defensive operations achieving 90%+ destruction rates. The September 2024 average reached approximately 75%, meaning three of every four launched drones were shot down before reaching targets.

But that 25% that penetrated defenses was sufficient to cause significant damage. If Russia launched 600 drones in a month and 150 reached their targets, striking power plants, fuel depots and civilian infrastructure, the campaign achieved strategic effects despite 75% losses. The mathematics favored attackers willing to accept high attrition rates, as long as production sustained the losses.

Tactical FPV drones showed inverse success rates. Ukrainian operators reported mission success rates drones, reaching and striking intended targets of 20-40% depending on electronic warfare environments. In heavily jammed sectors, success rates dropped to 10-15%. Against well-defended targets protected by air defense and electronic warfare, rates could fall below 10%. The implication was that achieving one successful strike required launching five to ten drones, sometimes more.

These ratios drove tactical planning. Battalion commanders calculated drone requirements based on expected attrition: destroying five enemy vehicles required launching 25-50 drones depending on target defenses. Suppressing an enemy position for an infantry assault required sustained drone presence, necessitating continuous launches to maintain pressure despite losses. Units with limited drone supplies had to choose targets carefully, concentrating precious resources against objectives where successful strikes would produce decisive effects.

The cost-exchange ratios told their own story. A $500 FPV drone that destroyed a $2 million Russian tank represented a 4,000:1 cost exchange favoring Ukraine. But that same drone, shot down by electronic warfare before reaching its target, represented a 100% loss with zero enemy damage. The variance between best-case and worst-case outcomes was enormous, making drone warfare simultaneously potentially decisive and potentially wasteful depending on execution.

Russian drones achieved similar success rates against Ukrainian targets. Ukrainian forces developed increasingly sophisticated counter-drone capabilities through 2024, combining electronic warfare, small arms fire, and dedicated counter-drone platforms. Russian FPV drones faced comparable attrition rates to Ukrainian drones, 70-80% in contested environments. The industrial drone war had become symmetrical, with both sides producing vast quantities, deploying them aggressively, and accepting high losses as the cost of achieving periodic successes.

The Trend Lines

Beyond static figures, trend lines revealed how the war was evolving. Month-to-month comparisons showed acceleration, deceleration, and shifts in strategic emphasis that single snapshots missed.

The comparison between August and September 2024 revealed significant escalation. Russian drone launches increased 39% from August to September, while missile launches increased 20%. The differential suggested Russian strategy was doubling down on drones as the primary strike weapon, with missiles reserved for specialized targets requiring precision or overwhelming destructive power that drones couldn't deliver.

But the longer trend showed more complex patterns. Russian drone launches had actually peaked in July 2024 at over 6,000 for the month before declining slightly in August and rebounding in September. The fluctuations reflected production bottlenecks, strategic pauses to build stockpiles for major offensives, and Ukrainian counter-attacks against drone production facilities and storage sites. The war wasn't a steady escalation but a series of surges and consolidations as both sides strained against industrial and logistical constraints.

Ukrainian production trends showed steadier growth. From producing perhaps 5,000 drones monthly in early 2023, Ukraine's production grew to 50,000 by mid-2023, 100,000 by early 2024, and 150,000+ by mid-2024. The growth curve was exponential through 2023, then logarithmic through 2024 as production approached practical limits imposed by component availability, manufacturing capacity, and skilled labor supply.

Looking forward, the announced expansion to 4 million drones annually represented another exponential leap, from 1.6 million to 4 million, a 150% increase. If achieved, this would put Ukraine on track to produce approximately 330,000 drones monthly, more than double current rates. The implications were staggering: over 10,000 drones daily, enough to sustain multiple simultaneous large-scale operations with reserves for replacement and stockpiling.

Russian production trends suggested similar expansion plans. Intelligence indicated new Shahed production facilities under construction, with completion expected in late 2024 or early 2025. Projected production rates of 600-700 Shaheds monthly by mid-2025 would provide Russia sustained saturation strike capabilities against Ukrainian infrastructure. FPV drone production was reportedly expanding as well, with the Russian defense industry aiming for 2 million units annually by 2025.

The competing expansion plans set up 2025 as a test of industrial endurance. Whichever side could produce faster, deploy more effectively, and sustain operations despite attrition would gain decisive advantages. The war was becoming explicitly industrial, victory would flow to whoever could manufacture more, faster, and sustain the grinding consumption that characterized drone warfare at scale.

The Economics

Behind the numbers lay economics, the financial costs of producing, deploying, and defending against millions of drones. The economic dimension of drone warfare was reshaping how nations thought about defense spending and military procurement.

At an average cost of $1,540 per drone (derived from Ukraine's $2.47 billion contract for 1.6 million drones), Ukraine was spending approximately $231 million monthly on drone production at October 2024 rates. Expanding to 4 million drones annually would require approximately $6.2 billion in annual spending roughly 8-10% of Ukraine's entire defense budget devoted to a single weapons category.

But this investment generated extraordinary combat power per dollar. A Ukrainian infantry brigade equipped with 1,000 drones monthly could conduct hundreds of strikes against enemy vehicles, fortifications, and personnel for approximately $1.5 million less than the cost of a single main battle tank. The return on investment, measured in enemy casualties and destroyed equipment, was unmatched by any conventional weapons system.

Russian economics showed similar patterns. If Russia produced 1.5 million FPV drones at comparable unit costs, annual spending approached $2.3 billion. Adding Shahed production estimated at $20,000-30,000 per unit for domestic Russian-built variants contributed another $100-150 million annually. Total Russian drone spending likely exceeded $3 billion annually by late 2024, a significant but manageable fraction of Russia's defense budget exceeding $80 billion.

The air defense economics were less favorable for Ukraine. A single Patriot missile cost approximately $4 million. Ukraine fired dozens monthly against Russian drones and missiles. Even cheaper systems like NASAMs missiles ($1 million each) or IRIS-T missiles ($400,000-500,000) imposed heavy financial burdens when employed against drone swarms. The economic exchange rates heavily favored attackers, Russia spent $20,000 to force Ukraine to spend $400,000 or more in defense, a 20:1 cost asymmetry.

Western military aid partially offset these costs, with the United States, European Union, and individual NATO nations providing air defense missiles and financial support for procurement. But the subsidization couldn't continue indefinitely. Western publics grew restive about bearing costs for defending Ukraine against attacks that showed no sign of stopping. The economics of drone warfare threatened to exhaust Western assistance before exhausting Russian capabilities.

The broader economic impacts were equally significant. Infrastructure destruction from successful drone strikes imposed costs measured in billions due to lost productivity, reconstruction expenses, economic disruption, and reduced GDP growth. Even unsuccessful attacks imposed costs through air defense expenditure, disrupted sleep reducing workforce productivity, and economic uncertainty deterring investment. Russia was waging economic warfare through drones, imposing costs that exceeded the price of the drones themselves by orders of magnitude.

The Human Arithmetic

Numbers quantified human costs as well, though imperfectly and incompletely. Casualty figures remained closely guarded by both sides, but fragmentary evidence suggested drones were causing increasing proportions of combat casualties.

Ukrainian military hospitals reported that drone-inflicted injuries constituted approximately 30-40% of combat casualties by late 2024, up from perhaps 10-15% in 2023. The increase reflected both the proliferation of drones and their growing lethality as warheads and tactics improved. Russian forces reportedly experienced similar patterns, with drones causing substantial fractions of their casualties.

Civilian casualties from drone strikes were better documented. Ukrainian authorities confirmed hundreds of civilians killed by Russian drone and missile attacks during 2024, with thousands more injured. The October 2024 figures showed 89 civilian deaths attributed to drone strikes specifically, not counting missile attacks. Each death represented individual tragedy—families destroyed, potential unrealized, loss multiplied across communities.

The psychological toll proved even harder to quantify but no less real. Mental health professionals estimated that millions of Ukrainians suffered from anxiety, depression, or post-traumatic stress related to constant aerial bombardment. Sleep deprivation affected entire populations. The cumulative psychological burden would echo for generations, affecting millions who survived physically but carried invisible wounds.

Operator casualties, drone pilots killed by counter-battery fire, targeted strikes, or accidents remained largely uncounted publicly but represented real losses. Elite drone units suffered attrition that while lower than infantry rates, was nonetheless significant. The psychological casualties, operators traumatized by the intimate killing that drone warfare entailed were higher still, though hidden from official casualty statistics.

The Logistics Challenge

Numbers also revealed the immense logistical challenge of sustaining drone warfare at scale. Producing 150,000 drones monthly meant sourcing millions of individual components, assembling them into functioning systems, testing for reliability, and distributing to hundreds of units across a thousand-kilometer front. The logistics were staggering.

Battery logistics alone posed enormous challenges. Each FPV drone required 1-2 lithium polymer batteries. At 150,000 drones monthly, Ukraine needed 200,000-300,000 batteries monthly just for new production, plus additional batteries for recharging existing drones. Batteries wore out after dozens of charge cycles, requiring constant replacement. The logistics of procuring, storing, charging, distributing, and disposing of millions of batteries was unprecedented in military operations.

Component logistics multiplied complexity. Motors, electronic speed controllers, flight controllers, cameras, video transmitters, frames and propellers, each drone contained dozens of specialized components sourced globally. Supply chain disruptions could halt production. Chinese export restrictions threatened critical components. Shipping delays cascaded through manufacturing schedules. Managing these supply chains under wartime conditions required sophisticated logistics usually associated with major automotive or electronics manufacturing, not military operations.

Distribution logistics challenged military organizations designed for conventional warfare. A battalion might receive thousands of drones monthly requiring storage, inventory tracking, distribution to subordinate units, training on new variants, and management of parts and maintenance supplies. Units without experience in retail-scale logistics suddenly needed warehouse management systems, inventory software, and specialist personnel.

Ammunition for drones, the grenades, shaped charges, and explosives that turned FPV platforms into weapons posed their own logistical challenges. Storing thousands of small explosive devices required specialized facilities meeting safety standards. Distribution had to balance availability with security. Tracking munitions through to employment prevented loss or misuse. The logistics resembled ammunition supply for small arms more than traditional military aviation.

The Strategic Implications

The numbers, taken together, painted a picture of warfare transformed. Annual production of 4 million drones by a single nation would exceed global military drone inventories before the war by two orders of magnitude. The scale represented not incremental change but revolutionary transformation in how wars were fought and sustained.

The arithmetic of attrition favored mass over sophistication. A single $30 million F-35 stealth fighter represented 60,000 FPV drones. The fighter could conduct perhaps 100-200 strike sorties over its service life before maintenance, wear, or combat losses ended its operations. The 60,000 drones could conduct 60,000 individual strikes, each potentially as lethal against ground targets as the fighter's weapons. The math was crude but revealing, mass-produced cheap weapons were achieving effects that previously required exquisite and expensive systems.

The production race's outcome would shape the war's trajectory. If Ukraine achieved 4 million annual production while Russia remained at 1.5-2 million, Ukraine gained decisive numerical advantages that could translate to operational success. If Russia matched or exceeded Ukrainian expansion, the grinding attrition would continue indefinitely until one side's industrial base collapsed or willingness to sustain costs eroded.

The numbers also illuminated lessons for future conflicts. Any nation facing adversaries with substantial industrial capacity had to plan for drone warfare at massive scale. Militaries needed not just sophisticated drones but the industrial capacity to produce them by thousands or millions. Defense strategies had to account for saturation attacks that could overwhelm even sophisticated air defenses through sheer numbers.

Colonel Danilov, surveying the ever-changing figures on the Ministry of Defense display, summarized the numerical reality: "Every number here represents decisions, sacrifices, achievements, and costs. The 1.6 million drones we're producing represent tens of thousands of workers, billions in investment, entire industries created from nothing. The 5,636 Russian drones we've faced represent 5,636 times our air defenses worked to protect civilians. The numbers tell the story of this war better than any battle narrative. And the story they tell is that we're in a contest of wills backed by industrial capacity. Whoever can sustain the numbers wins."

As 2024 turned toward 2025, the numbers continued climbing. Production rates increased. Consumption accelerated. Casualties mounted. Infrastructure suffered. But the war continued, grinding forward, sustained by statistics that would have seemed impossible three years earlier. The numbers mattered because they quantified reality not the reality of heroic narratives or strategic theories, but the concrete, countable, undeniable reality of industrial warfare in the drone age. And those numbers were staggering.

Chapter 14

The China Factor

The email arrived in Pavlo Kashchuk's inbox on a Tuesday morning in March 2024, its subject line reading simply "Order #47829 - Shipping Delay." Pavlo, procurement manager for a mid-sized Ukrainian drone manufacturer in Lviv, felt his stomach tighten. He opened the message already knowing what it would say. His supplier in Shenzhen, one of dozens of Chinese companies that had quietly supported Ukrainian drone production for two years, was informing him that DJI components could no longer be shipped to Ukrainian addresses. The restrictions had finally come.

"We are very sorry for this inconvenience," the email read in slightly awkward English. "Chinese government has implemented new export control regulations. DJI products cannot be exported to conflict zones. Your payment will be refunded within 7-10 business days. We hope for your understanding."

Pavlo didn't need to read further. He pulled up his spreadsheets showing component inventory and upcoming production schedules. Without DJI flight controllers, cameras, and motors, his company would exhaust existing stocks within three weeks. After that, production would halt unless alternatives could be found. He was far from alone across Ukraine, dozens of manufacturers were receiving similar messages, all pointing to the same reality: China, the indispensable supplier of drone components to both sides, was tightening its grip on the supply chain.

The China factor in the Ukraine drone war represented one of the conflict's most complex and consequential dimensions. While Russia and Ukraine fought on the battlefield, China occupied a different position, not quite neutral, not quite aligned, but absolutely central to both sides' ability to wage drone warfare. Chinese companies manufactured the motors, flight controllers, cameras, batteries, and countless other components that enabled the drone war. Chinese export policies could strengthen or strangle either combatant's capabilities. And Chinese strategic calculations about the war's trajectory and implications would shape decisions that affected the conflict as much as any battlefield engagement.

The Supply Chain Reality

To understand China's centrality, one must first grasp the global supply chain reality of drone manufacturing. The components that made FPV drones function came overwhelmingly from Chinese manufacturers, often from a concentrated geographic area around Shenzhen in Guangdong Province. This wasn't accidental, China had systematically developed dominance in consumer electronics manufacturing over decades, creating expertise, infrastructure, and economies of scale that competitors couldn't match.

DJI, the world's dominant consumer drone manufacturer, was merely the most visible example. Founded in 2006 by Frank Wang (Wang Tao), DJI had grown to control approximately 70% of the global consumer drone market by 2022. The company's Mavic series drones became ubiquitous in Ukrainian reconnaissance operations early in the war, prized for their reliability, sophisticated cameras, and ease of operation. Ukrainian forces employed thousands of DJI Mavics for artillery spotting, reconnaissance, and battle damage assessment.

But DJI represented just the tip of the supply chain iceberg. The motors that powered FPV drones came primarily from Chinese manufacturers like T-Motor, BrotherHobby, and Emax. Flight controllers, the sophisticated circuit boards that stabilized drones and interpreted pilot commands came from companies like Holybro, Matek, and JHEMCU. Electronic speed controllers, which regulated motor power, came from dozens of specialized Chinese manufacturers. Batteries, cameras, video transmission systems, antennas are all predominantly Chinese in origin.

Even components ostensibly from other countries often traced back to Chinese manufacturing. A "Swiss-designed" flight controller was likely manufactured in Shenzhen under license. "American" motors might be Chinese-made using American intellectual property. The global supply chain for small electronics had become so thoroughly dominated by Chinese manufacturing that avoiding Chinese components was practically impossible for anyone building drones at scale.

This dependency created vulnerability for both Ukrainian and Russian drone programs. Either combatant could potentially be strangled by Chinese export restrictions. And China, recognizing its leverage, began exercising that power with increasing assertiveness as the war progressed.

The Early Period: Benign Neglect

Through 2022 and into 2023, China's approach to Ukrainian drone procurement was characterized by what might be called benign neglect. Chinese companies sold components to Ukrainian buyers often through intermediaries to obscure the final destination without significant government interference. The business was lucrative, demand was enormous, and Chinese authorities appeared content to allow commerce to proceed.

Ukrainian procurement networks exploited this permissive environment aggressively. Volunteer organizations, private drone manufacturers, and even official government buyers established purchasing channels into China. Some worked through Hong Kong intermediaries who specialized in procurement for sensitive destinations. Others used sympathetic diaspora Chinese who could navigate both the technical requirements and the legal complexities. Still others simply ordered components as individual consumers or small businesses, with shipments routing through third countries to obscure their ultimate destination.

The procurement was often improvised and creative. One Ukrainian volunteer network purchased motor components marketed for electric skateboards and modified them for drone use. Another discovered that certain flight controllers intended for agricultural drones could be reprogrammed for military applications. The ingenuity wasn't just technical but logistical, finding ways to move thousands of small packages from Shenzhen to Kyiv despite wartime disruptions, customs complications, and increasing Chinese awareness of their military end-use.

Chinese authorities were certainly aware that their components were flowing to Ukraine for military applications. The question was whether Beijing cared enough to intervene. Several factors encouraged permissiveness: the business generated revenue for Chinese companies; restricting exports might push Ukraine toward developing alternative suppliers that would compete with Chinese manufacturers long-term; and China's official position of neutrality in the conflict suggested avoiding actions that clearly favored one side.

Russia enjoyed similar access to Chinese components during this period, though through different channels. Russian procurement benefited from closer official ties between Beijing and Moscow, with Russian companies often purchasing directly rather than through intermediaries. But Russia also employed subterfuge, especially after Western sanctions complicated financial transactions and technology transfers. Russian procurement networks used shell companies, falsified end-use certificates, and routing through neutral third countries to acquire Chinese components despite increasingly comprehensive Western sanctions.

DJI: The Pivotal Company

No single company's policies affected the drone war more significantly than DJI's. The Shenzhen-based manufacturer's products had become standard equipment for both sides, used for reconnaissance, targeting, and battle damage assessment. The company found itself in an increasingly uncomfortable position as its consumer products became essential military equipment in a major war.

DJI officially maintained neutrality and opposed military use of its products. Company statements emphasized that DJI drones were intended for civilian applications, photography, agriculture, infrastructure inspection, and recreation. The company implemented various restrictions meant to prevent military use: geofencing that prevented drones from flying in conflict zones, software updates that disabled certain features, and policies against selling to entities known to be military organizations.

But the restrictions proved easily circumvented. Ukrainian operators reprogrammed DJI drones to bypass geofencing. They ran older firmware versions that lacked the latest restrictions. They purchased drones as civilians through retail channels, then handed them to military units. The company's attempts to prevent military use were well-intentioned but technically ineffective against determined users with time and expertise.

The political pressure on DJI intensified from multiple directions. Western governments, particularly the United States, pressured the company to stop any sales that might benefit Russian forces. Ukrainian officials and advocacy groups demanded that DJI halt Russian sales while continuing Ukrainian access. Russian authorities wanted assured access while being content if Ukraine was denied the same. DJI found itself in an impossible position, any choice would anger someone, and potentially everyone.

By early 2024, DJI's response became increasingly restrictive. The company halted direct sales to Ukrainian addresses and tightened enforcement of policies against resale to military users. Ukrainian procurement through intermediaries became more difficult as DJI worked with Chinese customs authorities to identify and block suspicious shipment patterns. The company was carefully calibrating its position, restricting enough to satisfy

Beijing's demands for reduced support to belligerents while not cutting off access entirely, which would damage the company's reputation and invite competition.

For Ukrainian forces, the DJI restrictions created operational problems. Units that had relied on Mavic drones for reconnaissance suddenly faced supply disruptions. Training programs built around DJI's user-friendly platforms had to adapt to alternative systems with different interfaces and capabilities. Most significantly, the restrictions demonstrated that dependence on foreign suppliers created strategic vulnerabilities that could be exploited by geopolitical rivals.

The March 2024 Export Restrictions

The informal restrictions that had developed through 2023 became official policy in March 2024 when Chinese authorities announced new export control regulations targeting drone components and technology destined for "conflict zones." The regulations didn't specifically name Ukraine or Russia, but their intent was clear China was asserting control over a supply chain that had operated in regulatory grey zones for two years.

The new rules required exporters to verify end-use for any components that could be employed in military drones. Shipments to addresses in Ukraine, Russia, or neighboring countries required special licenses and documentation proving civilian applications. Customs inspections intensified, with suspicious shipments held for detailed examination. Financial transactions for drone components faced additional scrutiny, with banks required to report potentially military-related purchases.

The immediate impact on Ukrainian manufacturers was severe. Companies that had relied on direct Chinese shipments found orders blocked or delayed. Prices increased as intermediaries demanded higher fees for assuming the risks of circumventing restrictions. Delivery times lengthened as components took circuitous routes through third countries. Some manufacturers reported lead times doubling from 2-3 weeks to 4-6 weeks, severely constraining their ability to scale production or respond to urgent demands.

Ukrainian manufacturers responded through multiple strategies. Some established subsidiary companies in third countries such as Poland, Turkey, and UAE became popular destinations that could purchase components ostensibly for local markets, then transship to Ukraine. Others pivoted to smaller Chinese suppliers less subject to government scrutiny, trading the reliability of established manufacturers for the flexibility of companies willing to bend rules. Still others increased stockpiling, building inventory buffers that could sustain production through supply disruptions.

The restrictions weren't uniformly applied, suggesting Chinese authorities were balancing multiple objectives. High-profile components like complete DJI drones faced stringent controls. But less visible items such as generic motors, basic electronic components, and bulk materials still flowed relatively freely. China was sending a message about its power to constrain the conflict while avoiding restrictions so comprehensive they might drive permanent shifts in supply chains away from Chinese manufacturers.

The Russian Advantage

While both Russia and Ukraine faced Chinese export restrictions, Russia enjoyed significant advantages in navigating them. The deeper diplomatic relationship between Moscow and Beijing created channels for Russian procurement that Ukraine couldn't access. Russian companies could more easily secure export licenses by framing purchases as civilian or through official government-to-government arrangements that circumvented commercial restrictions.

More significantly, Russia could offer China what Ukraine couldn't: advanced military technology, energy resources, and political alignment on geopolitical issues. The relationship wasn't purely transactional but reflected deeper strategic convergence between Moscow and Beijing in opposition to Western-led international order. China's willingness to support Russian procurement even if indirectly through relaxed enforcement reflected this broader alignment.

Evidence of Chinese components in Russian weapons systems was substantial and well-documented. Western intelligence agencies examining captured Russian Shahed drones found Chinese-manufactured electronics, motors, and navigation systems throughout. Orlan reconnaissance drones, which Russia produced domestically, incorporated Chinese cameras and transmission equipment. Even Russian FPV drones, assembled in Russian facilities, used predominantly Chinese components that had entered Russia despite sanctions and export controls.

The smuggling networks that sustained Russian procurement were sophisticated. Chinese components flowed to Russia through Central Asian republics, particularly Kazakhstan and Kyrgyzstan, which had become major transshipment hubs. Russian front companies in these countries purchased components ostensibly for local use, then trucked them across porous borders into Russia. Other routes ran through Turkey, UAE, and even India, any jurisdiction where enforcement was lax or circumventable.

Chinese authorities certainly knew about these smuggling networks. The question was whether Beijing chose to enforce restrictions aggressively or allowed violations that served larger strategic objectives. The evidence suggested selective enforcement, cracking down when necessary to maintain plausible deniability while tolerating violations that supported Russia without creating diplomatic complications for China.

Ukraine's Forced Pivot

The Chinese export restrictions, while painful, ultimately accelerated Ukrainian efforts toward supply chain independence that strategic planners had recognized as necessary. The restrictions demonstrated that relying on geopolitically-aligned suppliers created vulnerabilities that adversaries could exploit. Ukraine needed domestic alternatives or diversified sourcing that reduced dependence on any single nation.

The push toward complete domestication of FPV drone production became a strategic priority. Ukrainian manufacturers began developing domestic alternatives to Chinese components wherever possible. Motor manufacturing started at small scale, with Ukrainian engineering firms producing electric motors based on reversed-engineered Chinese designs. Flight controller development accelerated, with Ukrainian software engineers creating firmware and control systems that matched Chinese capabilities.

The effort wasn't just technical but economic. Domestic manufacturing cost more than importing Chinese components, Ukrainian labor and overhead exceeded Chinese prices even accounting for shipping and customs. But the strategic calculation had shifted: paying 20-30% more for Ukrainian-made components was worthwhile if it eliminated supply chain vulnerabilities that could halt production entirely during critical periods.

By late 2024, Ukraine achieved significant progress toward supply chain independence for FPV drones. Domestic manufacturers could produce frames, motors, electronic speed controllers, and many other components at scale. Flight controllers remained partially dependent on Chinese chips, but Ukrainian companies had developed systems that could function with alternative processors. Batteries, among the most difficult components to replicate, still came predominantly from China, but Ukrainian and European sources were expanding production.

The achievement of "complete domestication of FPV drone production," as Ukrainian officials described it, was aspirational rather than absolute. Some components remained Chinese, sourced through increasingly elaborate procurement networks. But the direction was clear, Ukraine was systematically reducing dependence on foreign suppliers who might restrict access for geopolitical reasons. The Chinese export restrictions had inadvertently strengthened the Ukrainian defense industrial base by forcing innovations that wouldn't have occurred absent external pressure.

What China Was Learning

While China's export restrictions and supply chain dominance affected both combatants directly, the broader strategic implications centered on what China was learning from observing the drone war. Chinese military strategists, defense planners, and industrial policy makers watched the conflict with intense interest, drawing lessons that would shape Chinese military development for decades.

The first and most obvious lesson was the effectiveness of mass drone warfare. Chinese military doctrine had emphasized advanced technology and sophisticated systems to be stealth fighters, advanced missiles, and space-based capabilities. The Ukraine war demonstrated that large numbers of cheap, simple drones could achieve strategic effects previously requiring expensive, sophisticated platforms. Chinese defense planning began incorporating this reality, with visible increases in funding for mass drone production and development of swarming technologies.

The second lesson concerned supply chain vulnerabilities. China observed how dependence on foreign components created strategic weaknesses for both Russia and Ukraine. But China also recognized its own vulnerabilities, Chinese military systems incorporated American chips, European sensors, and Japanese precision components. A major conflict with Western powers would cut China off from these supplies. The war accelerated Chinese efforts toward technological self-sufficiency and supply chain resilience, particularly in semiconductors and advanced electronics where Chinese industry lagged Western competitors.

The third lesson involved Taiwan specifically. Chinese planners studying the drone war considered implications for potential Taiwan contingencies. Defending Taiwan against invasion would require exactly the kind of mass drone warfare that Ukraine pioneered, using swarms of cheap drones to attack Chinese amphibious forces, naval vessels, and air defenses. The tactics Ukrainians developed for defensive drone warfare were directly applicable to Taiwan's defensive needs. This realization intensified Chinese interest in counter-drone technologies and forced reconsideration of invasion plans that hadn't adequately accounted for defensive drone swarms.

Fourth, China learned about Western military aid patterns and sustainability. The war demonstrated both Western willingness to provide substantial support to aligned nations and the limits of that support. Western aid was generous but not unlimited, subject to political cycles and public opinion in donor countries. Chinese planners concluded that Western support might falter in prolonged conflicts, especially if adversaries could impose high costs through attritional warfare. The lesson suggested that China could potentially outlast Western support for Taiwan or other conflicts if Beijing could sustain operations despite initial losses.

Fifth, Chinese military observers studied the electronic warfare dimensions intensely. The constant battle between drones and jamming systems, the innovations in fiber-optic guidance and GPS-denied navigation, the importance of electromagnetic spectrum control all of it informed Chinese development of electronic warfare capabilities and counter-systems. Chinese engineers reverse-engineered captured Ukrainian and Russian drones, studied their electronics, and incorporated lessons into Chinese military systems.

The Broader Geopolitical Calculations

China's approach to the drone war supply chain reflected larger geopolitical calculations that extended beyond Ukraine. Beijing was navigating competing interests: maintaining economic relationships with Europe while supporting Russia strategically; developing military capabilities informed by the war while avoiding actions that might trigger Western sanctions; preserving neutrality officially while tilting toward Russia practically.

The economic dimension was significant. China's trade with Europe exceeded $800 billion annually, far larger than Russia-China trade. European pressure to restrict military-applicable exports to Russia created dilemmas for Beijing. Completely ignoring European concerns risked economic retaliation. But completely acquiescing to European demands would undermine Russia and suggest Beijing couldn't protect strategic partners. China threaded this needle through selective enforcement and plausible deniability, cracking down visibly on high-profile violations while tolerating smuggling networks that sustained Russian procurement.

The relationship with Russia itself was complex. China valued Russia as a strategic partner, a fellow great power challenging the American-led international order, and an energy supplier providing resources for Chinese economic growth. But Beijing also recognized Russian weakness revealed by the war. Russia's military struggled against a smaller adversary despite its nuclear arsenal and residual Soviet military infrastructure. The performance raised questions about Russia's value as an ally and created opportunities for China to extract concessions from Moscow's weakened position.

China's official neutrality served multiple purposes. It preserved relationships with European countries and even Ukraine itself, China-Ukraine trade continued during the war, albeit at reduced levels. It avoided triggering secondary sanctions from Western countries that might target Chinese companies for supporting Russian military operations. And it maintained China's positioning as potential mediator in the conflict, a role Beijing periodically emphasized through diplomatic initiatives that went nowhere but reinforced China's self-image as a responsible stakeholder in international affairs.

The Long-Term Implications

The China factor in the Ukraine drone war foreshadowed broader trends in international security and defense industrial relationships. The war demonstrated that supply chain dependencies were strategic vulnerabilities in ways most nations hadn't fully appreciated. Dominance in manufacturing wasn't just economic advantage but potential military advantage, the ability to strengthen allies or weaken adversaries by controlling access to critical components.

For Ukraine, the lesson was clear: strategic autonomy required industrial self-sufficiency, at least in critical technologies. The push to domesticate drone production would likely extend to other military systems. Ukraine emerged from the war determined to build defense industrial capabilities that couldn't be switched off by foreign suppliers' geopolitical calculations.

For Russia, Chinese supply chains proved both enabler and constraint. Chinese components allowed Russia to sustain drone warfare despite Western sanctions. But dependence on Chinese goodwill created vulnerabilities that Beijing could exploit for strategic leverage. Russia was learning the same lesson Ukraine learned that dependence was weakness.

For China, the war provided real-world testing ground for technologies, tactics, and strategies that would inform Chinese military development for decades. The lessons learned from Ukrainian and Russian experiences would shape Chinese drone warfare capabilities, counter-drone technologies, and broader military modernization. And China's position as an indispensable supplier gave Beijing leverage over European security that would complicate Western efforts to constrain Chinese power.

For the West, the China factor highlighted uncomfortable realities about economic interdependence and security competition. Western economies depended on Chinese manufacturing for everything from consumer goods to defense components. Decoupling was increasingly discussed but remained enormously difficult and expensive. The drone war suggested that technological competition with China would be complicated by Chinese dominance in manufacturing supply chains that Western nations had willingly ceded over decades.

Pavlo Kashchuk, the Ukrainian procurement manager who received that March 2024 email about export restrictions, summarized the strategic reality: "China holds the keys to the drone war. They can open doors or close them for both sides. We learned we can't depend on them, we have to build our own capabilities. But for now, we're still playing their game, trying to get components through whatever channels work. It's a race, can we become independent before they cut us off completely? That race matters as much as what happens on the battlefield."

As 2024 progressed, the China factor continued shaping the drone war in ways both visible and hidden. Components flowed through increasingly complex networks. Chinese policies evolved in response to diplomatic pressures and strategic calculations. And in Beijing, analysts studied the war's lessons, drawing conclusions that would shape Chinese military power and international security for decades to come. The drone war wasn't just being fought between Russia and Ukraine, it was being watched, analyzed, and learned from by the nation that would likely dominate 21st-century manufacturing and increasingly challenge Western military and economic power. What China learned from Ukraine would matter long after the shooting stopped.

PART IV: TOTAL DRONE WAR (2025)

Chapter 15

Industrial Scale Achieved

The factory floor stretched nearly the length of two football fields, a converted automobile parts plant on the outskirts of Kharkiv that had been repurposed into something unprecedented in military history. Rows of assembly stations lined the space in precise geometry, each staffed by workers in blue coveralls performing specialized tasks. At the first stations, carbon fiber frames were mounted on fixtures. Down the line, motors were affixed to arms with pneumatic tools. Further along, flight controllers were soldered, tested, calibrated. Near the end, completed drones underwent flight testing in netted enclosures before being packed into foam-lined boxes for shipment to the front.

Kateryna Shevchenko, the facility manager, stood on an elevated platform overlooking the operation, a tablet in her hands displaying real-time production metrics. "We're running at 1,847 units today," she said, her voice barely audible over the hum of hundreds of motors being tested simultaneously. "By the end of shift, we'll hit 2,000. That's forty thousand per month from this facility alone. Two years ago, I was managing the production of automotive alternators. Now I'm producing weapons of war at a scale that would have been unimaginable."

By early 2025, Ukraine had achieved something that military historians would study for decades: the transformation of a nation's entire industrial base to support mass production of unmanned aerial systems at a scale unprecedented in warfare. What had begun in garages and apartments in 2022 had evolved into an industrial ecosystem capable of producing 2.5 million drones annually, equivalent to roughly 7,000 drones every single day. The achievement represented not just technical accomplishment but organizational, logistical, and societal mobilization that redefined what was possible in wartime manufacturing.

The Path to 2.5 Million

The journey from thousands to millions had been neither smooth nor linear. Ukrainian drone production in 2022 was measured in hundreds per month, conducted by volunteers in improvised workshops. By mid-2023, production had scaled to 5,000-7,000 monthly across dozens of small manufacturers. The end of 2023 saw 50,000-75,000 monthly production. The contracted 1.6 million units for 2024 announced in March of that year represented approximately 130,000 drones monthly. Now, in early 2025, production capacity had nearly doubled again to 2.5 million annually, or roughly 208,000 drones per month.

The scaling required solving problems at every level. Material procurement at this volume demanded supply chains that hadn't existed in Ukraine before the war. Carbon fiber suppliers who had served small aerospace and sporting goods markets suddenly needed to provide tons of material monthly. Battery manufacturers expanded production from thousands to hundreds of thousands of units. Electronic component importers who had handled containers now handled cargo ships full of motors, controllers, and circuit boards.

Manufacturing capacity expansion meant constructing or repurposing dozens of major facilities. TAF Drones alone operated seven major production facilities by early 2025, each capable of producing thousands of drones daily. Smaller manufacturers proliferated reliable estimates suggesting 100-150 companies producing drones at varying scales, from boutique manufacturers making dozens of specialized units daily to major players producing tens of thousands monthly.

Quality control at scale presented enormous challenges. When producing hundreds of drones, manufacturers could test each unit thoroughly. At thousands daily, comprehensive testing became impossible, statistical sampling and automated testing replaced individual verification. Ukrainian manufacturers developed testing regimes that balanced reliability against throughput. Every tenth drone underwent flight testing. All electronic components were tested automatically before installation. Finished drones were spot-checked for critical failures, with batches quarantined if defect rates exceeded acceptable thresholds.

The workforce expansion was equally remarkable. The drone industry employed an estimated 50,000-70,000 workers by early 2025, spanning assembly line workers, engineers, quality control inspectors, logistics personnel, and support staff. Training pipelines produced thousands of skilled workers monthly, technical schools offered accelerated programs in drone assembly and testing, manufacturers operated their own training facilities, and experienced workers commanded premiums for training new employees.

Standardization emerged gradually as manufacturers realized the efficiency gains from common components and processes. Industry working groups established informal standards for motor mounts, flight controller interfaces, and battery connectors. While complete standardization proved elusive as manufacturers guarded proprietary innovations, enough commonality emerged that drones from different manufacturers could use interchangeable parts for basic components. Front-line units benefited enormously, able to cannibalize damaged drones for spare parts regardless of manufacturer.

TAF Drones: The Flagship

No company better illustrated Ukrainian drone manufacturing at industrial scale than TAF Drones. Founded in early 2023 by a group of engineers with backgrounds in telecommunications and aerospace, TAF had grown from a small workshop producing dozens of drones weekly into Ukraine's largest manufacturer, producing 40,000 drones monthly by late 2024 and expanding toward 50,000 monthly by early 2025.

The company's trajectory reflected both the opportunities and challenges of scaling military manufacturing during wartime. TAF's initial products were crude but functional, basic FPV platforms assembled from readily available components. But the founders recognized that sustainable success required moving beyond assembly toward actual manufacturing. TAF invested in tooling, machinery, and engineering talent that allowed the company to produce components rather than merely assembling imported parts.

By 2024, TAF manufactured its own frames using carbon fiber molding equipment, produced custom flight controllers with Ukrainian-designed firmware, and even experimented with motor manufacturing. The vertical integration provided resilience against supply chain disruptions while improving quality control and reducing costs. TAF drones incorporated increasingly sophisticated features, encrypted video transmission, GPS-denied navigation systems, and modular payload mounts that allowed rapid reconfiguration for different mission types.

The company's production facilities epitomized industrial-scale drone manufacturing. The main Kharkiv facility employed 3,000 workers operating three shifts daily. Material flowed through the factory in carefully orchestrated sequences, carbon fiber and resin at one end, packaged drones at the other, with dozens of steps in between. The facility consumed 50 tons of carbon fiber monthly, 200,000 motors, 150,000 electronic speed controllers, and 300,000 batteries. The logistics of managing these inputs while maintaining continuous production required sophisticated planning systems more typical of automotive manufacturing than military production.

TAF's annual revenue exceeded $1 billion by late 2024, making it one of Ukraine's largest defense contractors despite existing for less than two years. The company employed over 5,000 people across all facilities and support functions. It operated research and development programs developing next-generation platforms, maintained testing facilities where new designs were evaluated, and ran training programs that certified operators on TAF products. The company had evolved from garage startup to major defense industrial enterprise in less than 24 months, a transformation that would have taken decades in peacetime.

The human story behind TAF's growth was equally remarkable. Oleksandr Yaroshenko, TAF's co-founder and chief engineer, had been designing telecommunications equipment before the war. "I thought I understood manufacturing," he reflected. "I designed circuit boards, managed the production of electronic systems. But this scaling to forty thousand units monthly, managing supply chains under wartime conditions, dealing with constant material shortages and power outages was something entirely different. We learned by doing, made every mistake possible, and somehow built something that works."

The Distributed Ecosystem

While companies like TAF Drones represented the visible pinnacle of Ukrainian drone manufacturing, they existed within a broader ecosystem of hundreds of smaller manufacturers, specialized suppliers, and innovation labs that collectively generated Ukraine's production capacity. This distributed model, often criticized for inefficiency and lack of coordination, proved remarkably resilient and innovative.

Mid-sized manufacturers like Wild Hornets, Escadrone, and Vyriy Drone each produced 5,000-15,000 drones monthly, focusing on specific niches or design philosophies. Wild Hornets specialized in high-performance FPV platforms for elite operators, using premium components and hand-tuned flight characteristics. Escadrone focused on reconnaissance platforms with extended endurance and sophisticated sensors. Vyriy developed modular designs that allowed rapid reconfiguration for different mission types. The specialization meant Ukrainian forces could select drones optimized for specific tasks rather than using one-size-fits-all platforms.

Smaller manufacturers producing hundreds to thousands monthly provided innovation and flexibility that larger companies couldn't match. These boutique producers experimented with novel designs, tested unconventional configurations, and rapidly incorporated feedback from front-line operators. Many maintained direct relationships with specific combat units, producing custom drones tailored to particular operational needs. One small manufacturer developed specialized urban warfare drones for a single brigade operating in Mariupol, incorporating features that larger manufacturers would only adopt months later after the innovations proved effective.

Component manufacturers formed critical links in the supply chain. Ukrainian companies producing carbon fiber frames, circuit boards, and wire harnesses supplied both large and small drone manufacturers. Battery assembly operations took Chinese cells and packaged them into configurations optimized for specific drone platforms. 3D printing facilities produced specialized parts, motor mounts, payload adapters, camera housings allowing manufacturers to customize designs without expensive tooling for small production runs.

The cottage industry of individual craftsmen and micro-manufacturers persisted even as industrial production scaled. Skilled technicians working from home workshops produced specialized components, hand-wound motors, custom antennas, ruggedized electronics that commanded premium prices. Some front-line units maintained relationships with these craftsmen, who could produce custom solutions for unique problems faster than large manufacturers could respond to special requests.

This distributed ecosystem created redundancy that protected against disruption. When Russian missiles struck the TAF Drones facility in Kyiv in August 2024, destroying one production line, overall Ukrainian production barely declined, other manufacturers absorbed orders, production shifted to TAF's other facilities, and within weeks the damaged facility was producing again at reduced capacity. No single strike could cripple Ukrainian drone production because the capacity was distributed across the country in facilities both large and small.

The Coalition Support

Ukraine's achievement of 2.5 million annual production capacity wasn't entirely self-financed. International support, formalized through the Ukraine Drone Coalition, provided critical resources that enabled scaling at the pace achieved. The Coalition, formed in early 2024 and including dozens of nations, pledged €2.75 billion specifically for drone acquisition and production support.

Denmark emerged as the Coalition's most generous contributor, committing €1.5 billion for 2025 alone, an extraordinary sum from a country of 5.8 million people. The Danish funding model emphasized supporting Ukrainian manufacturing rather than providing finished drones. Danish money funded production line expansions, tooling purchases, workforce training, and research and development. The approach built Ukrainian capacity rather than creating dependency on foreign supplies.

Other Coalition members contributed through various mechanisms. Poland provided production facilities and technical expertise, hosting Ukrainian manufacturers who established subsidiaries to bypass supply chain restrictions. The United Kingdom funded specific production programs, particularly for advanced reconnaissance platforms incorporating British sensors and communications equipment. Germany provided precision tooling and manufacturing equipment that improved quality while increasing production rates. The United States, while not officially part of the Coalition, provided substantial funding through military aid packages that Ukraine directed toward drone procurement.

The Coalition funding allowed Ukrainian manufacturers to make investments that pure market economics might not have justified during wartime uncertainty. TAF Drones used Coalition support to build automated testing facilities that improved quality while reducing labor costs. Smaller manufacturers accessed Coalition grants for purchasing advanced manufacturing equipment, CNC machines, automated soldering systems and industrial 3D printers that increased production capacity and quality. Research initiatives funded through Coalition support developed advanced capabilities that would be incorporated into future drone generations.

The model represented new thinking about military aid. Rather than providing weapons systems from donor countries' arsenals, the traditional approach the Coalition funded recipient country production of relatively simple weapons systems. This built sustainable capability, supported the recipient country's economy, and avoided depleting donor countries' own military stockpiles. The success of the drone coalition model suggested approaches for supporting other nations facing security threats.

The 1.5 Million FPV Milestone

The 1.5 million FPV drones manufactured in 2024 represented a specific achievement worth examining separately. FPV platforms, the kamikaze drones that had proven most effective in tactical combat constituted the majority of Ukrainian production. The 1.5 million figure meant Ukraine produced approximately 4,100 FPV drones daily throughout 2024, or roughly 170 every hour around the clock.

To contextualize this production rate: every 21 seconds, somewhere in Ukraine, a completed FPV drone rolled off an assembly line ready for combat. Over the course of a year, Ukrainian manufacturers produced more FPV drones than all military drones of all types that had existed worldwide at the war's start. The scale was unprecedented and represented a transformation of warfare itself.

The FPV drones varied widely in specifications and capabilities. The majority were relatively simple platforms, seven to ten-inch frames with basic flight controllers, analog video transmission, and straightforward radio control. These workhorses cost $400-600 each and served as expendable strike weapons against opportunity targets. More sophisticated variants incorporated digital video, encrypted control links, enhanced electronic warfare resistance, and greater range. These premium platforms cost $1,500-2,500 but offered capabilities that justified higher costs for critical missions.

The manufacturing process for FPV drones had become remarkably efficient through iterative refinement. Experienced workers could assemble a basic FPV drone from components in 35-45 minutes. Automated processes such as computer-controlled soldering, robotic testing systems and programmed firmware flashing reduced assembly time further while improving consistency. The fastest production lines achieved cycle times under 30 minutes from components to packaged drones ready for shipment.

Quality metrics showed steady improvement despite accelerating production rates. Early 2023 FPV drones had failure rates approaching 15-20%, roughly one in six drones would malfunction before or during their mission. By 2024, failure rates for major manufacturers had declined to 5-8%, with premium manufacturers achieving 3-5%. The improvement came from better components, refined assembly processes, more comprehensive testing, and accumulated experience about which shortcuts were acceptable and which caused problems.

The human cost of producing 1.5 million drones was substantial. Estimates suggested each drone required approximately 2-3 person-hours of labor considering all steps from component preparation through final packaging. That meant 3-4.5 million person-hours of labor, equivalent to roughly 2,000 workers employed full-time throughout the year just for assembly. Adding support functions such as logistics, quality control, management and maintenance pushed total employment to perhaps 30,000-40,000 workers directly engaged in FPV drone production.

The Logistics Challenge

Producing 2.5 million drones annually was remarkable; distributing them to hundreds of units across a thousand-kilometer front while maintaining inventory management, tracking usage, and ensuring availability where needed and when needed posed challenges of equal magnitude. Ukrainian military logistics had to adapt to managing material flows unlike anything in prior military experience.

The scale of distribution was staggering. If Ukraine maintained two weeks of operational reserve, a conservative estimate that represented roughly 100,000 drones in field inventory at any moment, distributed across several hundred units. Each week required distributing approximately 50,000 new drones to replace combat losses and consumption. The drones moved through supply chains from factories to regional depots to brigade logistics nodes to battalion supply points to individual combat units, with each step requiring inventory management, transportation, and security.

Transportation logistics were complicated by several factors. Drones were simultaneously fragile, requiring careful handling to prevent damage to delicate electronic components and bulky relative to their weight, making packaging space-intensive. A truck that might carry 20 tons of ammunition carried only a few thousand drones. Moving 50,000 drones weekly required dozens of dedicated transport runs, each potentially targeted by Russian strikes against logistics networks.

Inventory management required sophisticated software systems tracking serial numbers, specifications, and locations for hundreds of thousands of drones. The system needed to answer questions like: Which units have the lowest drone inventories? Where are specific drone variants needed for planned operations? Which shipments are in transit? Which manufacturers have completed orders ready for pickup? Ukrainian logistics corps implemented commercial inventory management software adapted for military use, creating visibility into supply chains that enabled responsive distribution.

Battery logistics presented special challenges. Each drone required 1-2 batteries, meaning 2.5 million drones annually required 3-5 million batteries. Batteries degraded with use, typically becoming unreliable after 30-50 charge cycles requiring constant replacement. Field units needed charging infrastructure capable of servicing thousands of batteries daily. Used batteries required disposal through proper channels to prevent environmental damage and recover valuable materials. The logistics of battery management rivaled the drones themselves in complexity.

Maintenance and repair logistics emerged as production scaled. Early in the war, damaged drones were simply discarded, they were too cheap and too scarce to justify repair efforts. But at 2.5 million units annually with unit costs of $500-2,000, even modest repair rates saved significant money. Forward maintenance facilities staffed by trained technicians could repair drones damaged in transport or suffering minor malfunctions, returning them to service at fraction of replacement cost. The balance between repair and replacement required sophisticated calculations about labor costs, parts availability, and opportunity costs of technician time.

The Innovation Paradox

Achieving industrial scale created an innovation paradox. The processes, standardization, and scale economies that enabled production of 2.5 million drones annually worked against the rapid innovation that had characterized Ukraine's drone development. Manufacturers couldn't change designs weekly when production lines were optimized for specific configurations. Component standardization reduced flexibility to experiment with novel approaches. The focus on throughput and cost efficiency discouraged risky innovations that might improve performance but could disrupt production.

Yet innovation continued, albeit in different forms. Rather than constant design revolution, innovation became evolutionary, incremental improvements incorporated into existing platforms during periodic refresh cycles. Manufacturers established formal processes for evaluating innovations: small-scale prototype production, field testing with cooperative units, evaluation of performance and reliability, followed by gradual rollout if results justified adoption. The process was slower than the anarchic innovation of 2022-2023 but more systematic and scalable.

Research and development budgets grew as manufacturers' revenue increased. TAF Drones operated a dedicated R&D facility employing 200 engineers working on next-generation platforms, advanced sensors, improved electronic warfare resistance, and autonomous capabilities. The company tested dozens of experimental designs, most of which never reached production but informed incremental improvements to production platforms. The R&D investment represented long-term thinking, building capabilities for future conflicts, not just optimizing for current needs.

Some innovation moved outside major manufacturers to specialized labs and university research programs. Coalition funding supported academic research into drone technologies, creating knowledge that manufacturers could license and incorporate. Ukrainian technical universities established drone warfare research programs, attracting talented students who might otherwise have left the country. The research ecosystem would benefit Ukraine long after the war ended, building technical knowledge and human capital that could support civilian industries.

The Human Capital

Behind the statistics and industrial achievements were people, the workers, engineers, managers, and support personnel who transformed Ukraine's drone industry from nothing to global leader in 30 months. Their stories revealed both the opportunities and costs of this transformation.

Yuliya Kravchenko exemplified the journey many workers experienced. Before the war, she worked at a call center in Lviv for €600 monthly, a typical wage for entry-level service work. When TAF Drones opened a facility in Lviv in mid-2023, Yuliya applied for an assembly position offering €1,200 monthly, double her previous wage. She completed two weeks of training learning to solder circuit boards, test components, and assemble drone frames. Within six months, her skill and reliability earned promotion to line supervisor at €1,800 monthly. "The work is harder than the call center," she acknowledged. "Physically demanding, mentally tiring, and knowing what you're building will kill people isn't easy. But I'm contributing to defense, earning enough to support my family, and learning skills that will matter after the war ends."

Engineers commanded even higher premiums. Experienced electronics engineers could earn €3,000-5,000 monthly at major manufacturers, salaries comparable to Western European levels and extraordinary by Ukrainian standards. The companies competed aggressively for talent, offering not just high wages but good working conditions, modern facilities, and meaningful work. Brain drain, the emigration of skilled workers that had plagued Ukraine for decades reversed in the drone sector as engineers returned from abroad to participate in building something unprecedented.

The work took psychological tolls that weren't always visible in production statistics. Workers assembled weapons knowing each drone might kill Russian soldiers or might miss and kill civilians. Some workers found meaning in direct contribution to national defense. Others struggled with complicity in killing, even enemy combatants. Factory managers reported turnover rates of 20-30% annually, with mental health and moral discomfort cited alongside typical reasons for departure like better opportunities elsewhere.

Safety concerns were persistent. Working with lithium batteries, carbon fiber dust, and energized electronics created hazards that required careful management. Several serious accidents occurred, battery fires, machinery injuries, chemical exposures from adhesives and resins. Companies improved safety protocols iteratively, but the pressure to maximize production sometimes conflicted with safety measures that reduced throughput. The tension between production demands and worker safety was constant and not always resolved in workers' favor.

Looking Forward

By early 2025, Ukraine's achievement of 2.5 million annual drone production capacity represented both arrival and beginning. Arrival because Ukraine had reached industrial scale that established it as global leader in tactical UAV production. Beginning because plans already existed for further expansion, technological advancement, and organizational refinement.

The announced goal of 4 million drones annually would require another doubling of current production, a challenge given that the easier scaling had already occurred. Moving from 2.5 million to 4 million would require not just more of the same but solving new problems: finding additional skilled workers in tight labor market, securing sufficient raw materials as global supply chains strained to meet demand, and managing even larger and more complex logistics networks.

Technological advancement would continue pushing beyond current platforms. Artificial intelligence and autonomous systems represented the obvious next frontier, with several manufacturers developing prototype autonomous drones requiring minimal human control. Longer range, greater payloads, improved electronic warfare resistance, and better sensors would distinguish next-generation platforms from current workhorses. The innovation pipeline remained full despite production scaling.

Organizational evolution continued as the industry matured. Industry associations established standards, advocated for favorable policies, and coordinated with military customers. Labor unions emerged, organizing workers who wanted collective bargaining power. Regulatory frameworks developed to govern an industry that hadn't existed three years earlier. The anarchic early days were giving way to institutional structures that would persist beyond the war.

Kateryna Shevchenko, the factory manager watching 2,000 drones roll off her production line daily, captured both the achievement and its surreal nature: "Sometimes I step back and really look at what we're doing, producing thousands of flying weapons every day, in a factory that made car parts three years ago, with workers who were doing completely different jobs before the war. It's extraordinary. It's also horrifying. We've built something remarkable that I hope humanity never needs to build again. But we built it. And that matters."

The industrial scale achieved by early 2025 would define Ukraine's war effort and influence global military development for decades. The numbers 2.5 million drones annually, 40,000 from single manufacturers monthly,

€2.75 billion in international support told a story of transformation that changed warfare itself. Industrial capacity had become combat power in ways that previous generations of military planners had never imagined. Ukraine had proven that determined nations could build world-leading capabilities in emerging technologies faster than established powers could adapt. The lesson would not be forgotten.

Chapter 16

Deep Strike Capabilities

The first sign that something had changed came at 3:47 AM on August 12, 2025, when alarms screamed at the Ryazan Oil Refinery, 196 kilometers southeast of Moscow. Security cameras captured the moment of impact, a dark shape descending from the night sky, moving with purposeful precision toward Tank Farm 3. The explosion that followed engulfed three storage tanks in flames that would burn for eighteen hours, destroying 50,000 tons of refined petroleum products and sending a pillar of black smoke visible from space.

By dawn, Russian authorities were still struggling to comprehend what had occurred. The drone, Ukrainian certainly had flown nearly 500 kilometers from the border to strike deep inside Russia proper. It wasn't the first such attack, but it represented an escalation in both range and precision that marked a new phase of the war. Ukraine had developed strategic bombing capabilities not through aircraft that could be shot down, but through unmanned systems that could penetrate hundreds of kilometers into enemy territory, strike critical targets, and reshape the strategic balance of the conflict.

The evolution of Ukrainian long-range strike drones from experimental systems to operational weapons revolutionized warfare in ways that conventional military planning hadn't anticipated. Strategic bombing, the ability to attack an adversary's industrial base, logistics networks, and resource production deep behind front lines had traditionally required expensive bomber aircraft, cruise missiles costing millions of dollars each, or ballistic missiles available only to advanced military powers. Ukraine demonstrated that similar effects could be achieved with drones costing tens of thousands of dollars, produced by small manufacturers, and operated by small teams with minimal infrastructure.

The Strategic Imperative

Ukraine's development of deep strike capabilities emerged from strategic necessity rather than technological preference. By mid-2024, the ground war had calcified into an attritional stalemate where neither side could achieve breakthrough despite enormous casualties and material consumption. Russian forces, despite setbacks, maintained pressure along multiple axes. Their capacity to sustain operations depended on logistics networks, industrial production, and energy resources largely invulnerable to Ukrainian conventional forces whose artillery and tactical aviation couldn't reach targets deep inside Russia.

The asymmetry was stark and painful. Russia routinely struck targets throughout Ukraine such as power plants, factories, transportation hubs, government buildings using missiles and drones launched from Russian territory. Ukraine, constrained by Western restrictions on using provided weapons against targets inside Russia and lacking indigenous long-range strike capabilities, could respond only by defending against attacks while its critical infrastructure was systematically destroyed.

This imbalance wasn't merely tactical, it was existential. If Russia could attack Ukraine's industrial base with impunity while Ukraine couldn't reciprocate, the attrition war would inevitably favor Russia's larger economy and deeper resources. Ukraine needed capabilities to strike Russian territory, imposing costs that complicated Russian operations and potentially shifting calculations about whether continued war was worthwhile.

The strategic targets were obvious. Russian oil refineries processed crude oil into gasoline, diesel, and aviation fuel that sustained military operations and generated export revenue funding the war. Logistics hubs concentrated military supplies before distribution to front-line forces. Military-industrial facilities manufactured weapons, ammunition, and equipment. Railway networks moved forces and supplies across Russia's vast territory. Transportation infrastructure, bridges, ports and marshalling yards channeled resources toward the war effort. All were legitimate military targets that could significantly degrade Russian war-making capacity if successfully attacked.

But reaching these targets, hundreds of kilometers inside Russia, required capabilities Ukraine didn't possess in 2022 or early 2023. The solution would come through Ukrainian innovation in long-range unmanned systems, platforms that extended tactical drone technology into the strategic domain.

Evolution of Long-Range Platforms

Ukrainian long-range strike drones evolved through several generations, each extending range, improving reliability, and enhancing precision. The development wasn't linear or centrally planned but emerged from multiple parallel efforts by different manufacturers, military units, and volunteer groups who recognized the strategic imperative and worked toward solutions.

The earliest long-range strikes in 2023 used modified commercial drones pushed beyond their designed capabilities. Ukrainian engineers would strip unnecessary components, install oversized fuel tanks or additional batteries, program waypoint navigation, and launch these improvised platforms toward distant targets. Success rates were abysmal, perhaps 10-20% reached their targets, with most failing due to mechanical problems, running out of fuel, or being shot down. But even sporadic successful strikes demonstrated the concept's potential.

By late 2023, dedicated long-range platforms began emerging from Ukrainian manufacturers. These weren't modifications but purpose-built systems designed for deep penetration missions. Typical configurations used fixed-wing designs for fuel efficiency, gasoline engines for extended endurance, and sophisticated navigation systems combining GPS, inertial guidance, and terrain-following algorithms. Payloads ranged from 20-50 kilograms of explosives, enough to cause significant damage to industrial targets but small enough to keep the drone light and fuel-efficient.

The "Queen Hornet" platform, developed by Wild Hornets in collaboration with military intelligence, exemplified the type. With a wingspan of 3.5 meters and length of 2.8 meters, the Queen Hornet resembled a large model airplane more than a military weapon. Its pusher-propeller configuration powered by a small gasoline engine could sustain flight for 10-12 hours, providing a range exceeding 1,000 kilometers in ideal conditions. The drone carried 30 kilograms of explosives in a shaped charge optimized for penetrating industrial structures. Unit cost was approximately $35,000-50,000, expensive by FPV standards but trivial compared to cruise missiles costing millions.

Navigation represented the critical technical challenge. Flying hundreds of kilometers, often through Russian air defense zones, required sophisticated guidance that could function despite GPS jamming or denial. Ukrainian engineers developed hybrid navigation systems using GPS when available, switching to inertial navigation during jamming, and incorporating terrain-matching algorithms that compared radar altimeter data with pre-loaded terrain maps to verify position. The systems weren't perfect, navigation errors of several hundred meters were common but provided sufficient accuracy to hit large fixed targets like refineries, power plants, or factory complexes.

Stealth wasn't achieved through exotic materials or radar-absorbent coatings, Ukraine lacked the technology and resources for such sophistication but through operational tactics. The drones flew at low altitudes, using terrain masking to avoid radar detection. They launched at night to reduce visual detection. Flight paths avoided known air defense concentrations. Some missions used multiple drones approaching from different directions simultaneously, overwhelming defensive systems through saturation rather than stealth. The approach was crude but effective enough that significant numbers of drones reached their targets.

By summer 2025, multiple manufacturers were producing long-range strike platforms with improving capabilities. Ranges extended to 1,200-1,500 kilometers for specialized variants. Payloads increased to 40-60 kilograms as engines and aerodynamics improved. Navigation accuracy tightened through better sensors and algorithms. Production costs declined through manufacturing refinement and economies of scale. What had been experimental systems in 2023 were becoming mature weapons by 2025.

The August 2025 Campaign

The systematic campaign of deep strikes that began in August 2025 demonstrated how far Ukrainian capabilities had progressed. Rather than sporadic opportunistic attacks, Ukraine launched coordinated strikes against Russian oil infrastructure with frequency and precision that shocked Russian authorities and Western observers alike.

The targeting priorities reflected careful strategic planning. Oil refineries topped the list, Russia operated approximately 30 major refineries processing crude oil into refined products. These facilities were concentrated in specific regions: the Volga-Urals area, Western Siberia, and the Moscow region. Each refinery represented a critical node in Russian energy production, and each was vulnerable to drone strikes despite security measures and air defenses.

The Ryazan refinery strike on August 12 was followed by attacks on the Yaroslavl refinery on August 15, the Nizhny Novgorod facility on August 18, and the Samara refinery on August 22. Each attack followed similar patterns: drones launched in darkness, approaching targets at low altitude, striking tank farms or processing units that caused spectacular fires and significant disruption to production. Russian air defenses destroyed some attacking drones, but enough penetrated to achieve their objectives.

The cumulative effect was substantial. Russian oil refining capacity declined an estimated 8-12% through the August-September 2025 period, forcing imports of refined products and complicating logistics for military operations that depended on steady fuel supplies. Market prices for Russian oil products increased as refined output couldn't meet domestic demand. The attacks imposed costs measured in billions of rubles while Ukrainian investment was measured in millions.

Beyond refineries, Ukrainian drones struck ammunition depots, causing catastrophic explosions that destroyed stocks accumulated over months. Railway bridges and marshalling yards suffered attacks that disrupted logistics flows. Military airfields lost aircraft to drone strikes that penetrated air defenses and destroyed planes on the ground, a bitter irony given that airfields were supposed to be among the most heavily defended locations. Each successful strike demonstrated both Ukrainian capability and Russian vulnerability.

The operations weren't without failures. Many drones were shot down by Russian air defenses, crashed due to mechanical failures, or missed targets because of navigation errors. Russian electronic warfare disrupted some missions completely. But the success rate estimated at 40-60% of launched drones achieving mission objectives was sufficient to sustain the campaign. Ukraine could afford to lose half the drones launched if the remainder struck targets imposing disproportionate costs on Russian operations.

The psychological impact complemented the physical damage. Russian civilians, previously insulated from direct war effects beyond economic sanctions and mobilization, now experienced attacks deep inside Russia. Moscow residents watched smoke plumes from struck refineries. News of drone strikes on industrial facilities near population centers created anxiety that war might come to Russian cities as it had come to Ukrainian cities. The attacks punctured Russian propaganda narratives about the war being under control and Ukraine being on the verge of defeat.

Russia's Reciprocal Campaign

Russia, of course, possessed its own long-range strike capabilities and employed them with equal ruthlessness. Russian long-range drones and missiles struck Ukrainian railways, attempting to disrupt the logistics networks that sustained Ukrainian military operations and connected Ukraine to Western supply routes through Poland and Romania.

The targeting logic was clear: Ukrainian railways moved virtually everything, Western military aid from Polish border crossings, Ukrainian military units deploying between sectors, civilians evacuating threatened areas, and commercial goods that kept the economy functioning. The rail network, built during Soviet times when Ukraine was integrated into Russian economic space, remained the backbone of Ukrainian transportation. Disrupting it would cascade into operational and strategic problems affecting everything from military operations to civilian morale.

Russian strikes hit railway stations, switching yards, bridges, maintenance facilities, and rolling stock. The attacks employed both expensive cruise missiles for the most critical nodes and cheaper Shahed drones for widespread harassment. The combination forced Ukrainian railway workers into dangerous conditions, repairing tracks under threat of additional strikes, operating trains that might be targeted, and maintaining systems under constant attack.

Ukrainian railways demonstrated remarkable resilience. Repair crews worked around the clock fixing damaged tracks, replacing destroyed switches, and rebuilding bombed facilities. Redundancy in the network meant that even significant damage could be routed around. The distributed nature of railway operations, dozens of critical junctions, hundreds of kilometers of track made comprehensive destruction impossible unless Russia was willing to expend enormous quantities of weapons.

But resilience didn't mean invulnerability. The constant attacks degraded capacity, trains moved more slowly through damaged sections, routing around destroyed infrastructure added time and distance to journeys, and the psychological toll on railway workers affected operations. Ukrainian logistics, while not paralyzed, operated under constraints that complicated military operations and economic activity.

The mutual deep strike campaigns created strategic dilemmas for both sides. Each successful strike justified the effort and resources invested. But neither side could achieve decisive effects through deep strikes alone, the economies were too large, the target sets too distributed, and the defenses too effective to allow complete destruction of critical infrastructure. The campaigns imposed costs and complicated operations but didn't substitute for success in ground combat that would determine the war's outcome.

Technical and Operational Challenges

Conducting deep strike operations at scale presented challenges beyond simply building capable drones. The entire operational chain, intelligence, targeting, mission planning, launch, in-flight control and battle damage assessment required capabilities that Ukraine had to develop from scratch.

Intelligence collection for targeting deep inside Russia was extraordinarily difficult. Satellite imagery could identify refineries and military facilities, but understanding vulnerable points within facilities, which tanks held refined products versus crude oil, which buildings housed critical equipment versus offices required detailed intelligence often unavailable. Ukrainian intelligence services recruited human sources, analyzed commercial satellite imagery, and pieced together information from open sources to build targeting databases. The process was laborious and imperfect, but sufficient to identify targets worth striking.

Mission planning required sophisticated analysis. Planners needed to chart flight paths avoiding Russian air defense concentrations while allowing terrain masking. They calculated fuel requirements accounting for winds, temperature, and altitude variations. They timed launches so drones would arrive at targets during optimal conditions, typically pre-dawn hours when visibility was lowest and defensive readiness potentially reduced. They planned for contingencies: what happens if a drone encounters unexpected jamming? If weather forces altitude changes? If the target has been moved or is more heavily defended than expected?

Launch operations presented their own challenges. Long-range drones were larger and more complex than FPV platforms, requiring dedicated launch facilities with runways or catapult systems. Launch sites needed to be positioned close enough to targets to be within drone range while far enough from front lines to avoid Russian strikes. Security was critical, Russian intelligence actively sought launch sites for targeting. Mobile launch teams operated from different locations for each mission, moving after launch to avoid counter-strikes.

In-flight control was minimal for most long-range strikes. The drones were essentially fire-and-forget weapons, programmed with waypoint navigation and launched toward targets without the ability for operators to intervene mid-flight. Some platforms incorporated limited command links allowing course corrections if communications remained intact, but heavy Russian electronic warfare often precluded reliable control beyond launch. The lack of control meant missions couldn't adapt to changing circumstances, if a target moved or defenses proved unexpectedly strong, the drone would strike anyway or crash trying.

Battle damage assessment required different intelligence collection efforts. Post-strike satellite imagery showed physical damage but couldn't always determine operational impacts. Was a burning refinery completely offline or operating at reduced capacity? How long would repairs take? Had the strike destroyed critical equipment or merely damaged less essential structures? The assessment challenges meant Ukrainian planners sometimes struck targets that had already been adequately damaged or missed targets that remained operational despite apparent destruction.

The Political and Diplomatic Dimensions

Ukrainian deep strikes into Russia created political and diplomatic complications that extended beyond military effectiveness. Western allies who provided weapons and financial support to Ukraine had explicitly restricted use of their systems against targets inside Russia, fearing escalation that might draw NATO into direct conflict. Ukrainian strikes using indigenous systems operated in different legal and political spaces, but still generated concern among allies worried about escalation dynamics.

The United States, Ukraine's most important supporter, maintained ambiguous positions on Ukrainian deep strikes. Officially, the U.S. neither encouraged nor prohibited Ukrainian operations inside Russia using Ukrainian-developed systems. Privately, American officials expressed both admiration for Ukrainian technical achievements and concern about strategic risks. Some U.S. analysts worried that strikes on Russian territory might provoke Russian retaliation, potentially including nuclear weapons use that could expand the war beyond Ukraine. Others argued that Ukrainian strikes imposed costs that made Russian victory less likely, serving Western interests in avoiding Russian success.

European allies were divided on the issue. Eastern European nations generally supported Ukrainian strikes as legitimate self-defense responses to Russian attacks on Ukrainian territory. Western European countries were more cautious, concerned about escalation and preferring negotiations to intensified combat. The divisions reflected broader European debates about how to approach the conflict, whether to support Ukraine toward military victory or pressure both sides toward negotiated settlement.

Russia, predictably, condemned Ukrainian strikes as terrorism and threatened retaliation. But Russian responses were constrained by the reality that Russia was already striking Ukrainian territory extensively. Moscow could threaten escalation, but to what? Russia was already employing most capabilities short of nuclear weapons. The deep strike campaigns demonstrated limits of Russian escalation dominance threats became less effective when adversaries called bluffs and continued operations despite threatened consequences.

The strikes also complicated potential negotiations. Russian negotiators insisted that any ceasefire or peace agreement must include Ukrainian commitments to cease attacks on Russian territory. Ukrainian negotiators countered that Ukraine would cease strikes when Russia ceased attacks on Ukraine, including strikes on civilian infrastructure that had caused far more damage than Ukrainian attacks on Russian military and industrial targets. The mutual deep strike campaigns became obstacles to negotiation while simultaneously being tools each side hoped would improve their negotiating position.

Strategic Effects and Limitations

By late 2025, enough evidence existed to assess the strategic effects of Ukrainian deep strike capabilities. The picture was complex, significant tactical successes that imposed real costs on Russian operations, but falling short of decisive strategic impact that would fundamentally alter the war's trajectory.

On the positive side from Ukrainian perspective, the strikes had successfully:

- Reduced Russian oil refining capacity by 8-12%, complicating fuel logistics and reducing export revenues
- Destroyed substantial ammunition stocks, forcing Russia to draw down reserves or reduce ammunition consumption at the front
- Demonstrated Ukrainian reach into Russian territory, puncturing narratives about Russian security and control
- Imposed costs that exceeded the resources Ukraine invested in the strikes by factors of 10-100
- Created domestic political pressure within Russia as citizens experienced war's consequences directly
- Provided asymmetric response to Russian strikes on Ukrainian infrastructure, achieving rough parity in mutual vulnerability

But the strikes had not achieved more ambitious objectives like:

- Forcing significant reallocation of Russian military resources from front lines to rear area defense (some reallocation occurred but was modest)
- Generating domestic Russian opposition to the war sufficient to pressure Putin toward negotiations
- Disrupting Russian logistics so severely that front-line operations were critically degraded
- Destroying sufficient industrial capacity to prevent Russian military production from sustaining operations
- Achieving strategic effect equivalent to conventional strategic bombing campaigns by manned aircraft

The limitations reflected both Russian adaptations and inherent constraints on drone strike effectiveness. Russia improved air defenses around critical infrastructure, making successful strikes more difficult and costly. Critical facilities were hardened with additional protection. Repair capabilities improved as Russia gained experience responding to strikes. And most fundamentally, Russia's size and resource depth meant that even successful strikes couldn't destroy enough capacity to cripple operations, destroyed refineries could be replaced by others, disrupted railways could be routed differently, and damaged facilities could be repaired or replaced.

The strategic lesson was nuanced: deep strike drones provided capabilities that Ukraine desperately needed and achieved effects that justified the investment. But drones weren't silver bullets that could substitute for success in ground combat or provide decisive strategic advantage. They were tools, valuable tools, but tools nonetheless that contributed to broader campaign plans but couldn't independently determine outcomes.

The Future of Deep Strike

Looking forward to late 2025, several trends seemed likely to shape future deep strike operations. Range would continue increasing as engines, aerodynamics, and fuel systems improved, platforms capable of 2,000+ kilometer ranges were under development. Payloads would grow as larger platforms became practical. Precision would improve through better navigation and possibly terminal guidance systems allowing strikes on smaller, more specific targets.

Swarm capabilities represented the next frontier. Rather than launching individual drones toward targets, future operations might employ dozens of drones launched simultaneously toward multiple targets or converging on single targets from multiple directions. Coordinated swarms could overwhelm air defenses, improve probability of successful strikes, and achieve effects impossible for individual drones.

Autonomous systems would gradually replace human-programmed navigation. Artificial intelligence could enable drones to identify and adapt to air defenses, select optimal approach routes in real-time, distinguish primary from secondary targets, and make terminal attack decisions without human control. The technology posed ethical and legal questions, but military pressure to improve effectiveness would likely drive adoption despite concerns.

Counter-drone capabilities would advance in parallel, creating ongoing technological competition between penetration and defense. Long-range interceptor drones, improved radar and electronic warfare systems, and directed energy weapons would all be developed to counter evolving strike platforms. The competition wouldn't end, it would continue indefinitely as offense and defense spiral upward in capability.

The normalization of deep strikes as routine military operations represented perhaps the most significant development. What began as an extraordinary achievement, Ukraine striking targets 500 kilometers inside Russia was becoming normal by late 2025. Future conflicts would almost certainly feature similar operations from the outset rather than developing them over years of combat. The genie was out of the bottle, and strategic bombing with drones had become part of modern warfare's standard toolkit.

Major Oleksandr Bondar, who commanded a Ukrainian long-range strike unit, reflected on the transformation: "Two years ago, we were experimenting with modified racing drones hoping to reach targets fifty kilometers away. Now we're hitting refineries five hundred kilometers inside Russia with purpose-built platforms we manufacture ourselves. The technology progressed faster than anyone imagined possible. The strategic implications are still unfolding. We've proven that even relatively small nations with modest resources can strike deep into adversary territory if they're willing to innovate. That changes deterrence calculations, alliance commitments, and strategic planning worldwide. We're not just fighting our war, we're demonstrating how future wars will be fought."

The deep strike capabilities that emerged from August 2025 onward redefined what was possible in modern warfare. Strategic bombing, once the exclusive domain of major powers with expensive bomber fleets or sophisticated missile arsenals, had been democratized through drone technology. The implications would echo long after the Russia-Ukraine conflict ended, shaping military development and strategic thinking for decades to come. Ukraine had proven that reaching deep into enemy territory required neither stealth bombers nor cruise missiles, just determination, ingenuity, and drones.

Chapter 17

Urban Warfare Evolution

The apartment building on Heroiv Pratsi Street in Kherson had seen better days. Shells had punctured its facade, windows gaped empty, and the top two floors were partially collapsed from artillery strikes. But people still lived there, elderly residents who refused evacuation, families with nowhere else to go, and a handful of Ukrainian soldiers using the building's basement as an observation post. On the morning of October 3, 2025, they would learn that the front line had extended far beyond what military maps suggested, penetrating directly into their living rooms and kitchens through a threat that arrived silently on rotors barely larger than a hand.

Mariya Tkachenko, 67, was preparing tea in her third-floor apartment when she heard the distinctive high-pitched whine approaching. She'd learned to recognize the sound over months of occupation and liberation, a Russian FPV drone hunting. She dropped the kettle and ran toward the hallway, but the drone was faster. It crashed through what remained of her window, detonated against the interior wall, and filled the apartment with shrapnel and flame. Mariya survived with severe injuries. Her neighbor, 72-year-old Petro Lysenko, did not.

The incident was neither unique nor particularly remarkable by late 2025 standards. Russian forces, unable to advance significantly on the ground against Ukrainian defenses, had increasingly employed long-range FPV drones to attack targets deep inside Ukrainian-controlled cities. The fiber-optic cables that made these drones immune to jamming allowed them to penetrate 15-20 kilometers beyond the nominal front lines, turning entire urban areas into contested battlespaces where nowhere was truly safe. Urban warfare had evolved beyond street-to-street combat between infantry forces into something more insidious, a constant aerial threat that dissolved the distinction between front line and rear area, combatant and civilian.

The Collapse of Safe Zones

Traditional military thinking about urban warfare distinguished between combat zones where active fighting occurred and rear areas where support functions, wounded treatment, and civilian life continued with relative safety. The paradigm assumed that only the forward edge of the battle area was truly dangerous, with risk declining rapidly with distance from front lines. Infantry might fight building-to-building in contested neighborhoods, but hospitals, command posts, and civilian areas several kilometers back were secure except from long-range artillery or occasional air strikes.

Drone warfare obliterated these comfortable distinctions. A Russian FPV drone launched from positions near the Dnipro River could reach targets throughout Kherson, a city of 280,000 before the war within 15-20 minutes. Fiber-optic guidance meant electronic warfare couldn't stop them. The drones flew low, using buildings for cover, making air defense difficult. And the operators, watching through high-definition video feeds, could select targets with precision, a Ukrainian soldier smoking a cigarette, a vehicle delivering supplies, civilians queuing for humanitarian aid.

Kherson exemplified the phenomenon but wasn't unique. Bakhmut during its months-long siege, Mariupol during its desperate defense, Avdiivka during Russian assaults, and dozens of smaller cities and towns experienced similar dissolution of safe zones. Russian drones reached deep into Ukrainian-controlled urban areas, while Ukrainian drones penetrated Russian-held cities. The mutual vulnerability transformed urban warfare into something qualitatively different from historical precedents.

The psychological impact on residents was profound and corrosive. During World War II's urban battles of Stalingrad, Berlin and Manila civilians could seek shelter in basements and wait for front lines to pass. Modern urban warfare offered no such refuge. A basement might protect against artillery but not against a drone deliberately flown into an opening to attack whoever sheltered inside. Staying indoors provided no safety, drones crashed through windows and doors. Movement attracted attention and drones hunted people visible on streets. The only truly safe response was to evacuate entirely, abandoning homes and communities to empty shells contested by forces using them as hunting grounds.

Evacuation itself became a target for drone attacks. Russian operators observed evacuation routes and targeted vehicles carrying civilians attempting to flee. Humanitarian corridors, theoretically protected under laws of war, became killing zones where drones struck buses, cars, and even people on foot. The tactic was ruthlessly pragmatic from a purely military perspective, forcing civilians to remain in contested areas complicated Ukrainian military operations but constituted clear war crimes that would eventually face international accounting.

Fiber-Optic Penetration

The fiber-optic drones introduced in spring 2024 fundamentally changed urban warfare dynamics by extending effective range and eliminating the electronic warfare protection that Ukrainian forces had relied upon. Traditional radio-controlled drones operating in urban environments faced multiple challenges: buildings blocked line-of-sight control signals, electromagnetic interference from civilian infrastructure disrupted communications, and Ukrainian electronic warfare created jamming zones protecting critical areas. Fiber-optic guidance bypassed all these limitations.

The operational pattern became grimly familiar. Russian drone teams would position themselves at forward positions, often in high-rise buildings or elevated terrain providing good views with fiber-optic FPV drones ready for launch. Reconnaissance drones or human observers would identify targets in Ukrainian-controlled areas: troop concentrations, command posts, supply vehicles, or simply clusters of people who might be soldiers. The fiber-optic drones would launch, paying out cable as they flew toward targets, guided by operators watching real-time video feeds that no amount of jamming could disrupt.

The fiber-optic cables, typically 10-15 kilometers in length, provided sufficient range to reach most areas within cities like Kherson, where Russian positions along the eastern bank of the Dnipro River sat roughly 5-10 kilometers from much of the city center. For cities like Bakhmut or Avdiivka that had Russian forces on multiple sides, the vulnerability was even greater, drones could approach from any direction where Russian positions existed within cable range.

Ukrainian forces adapted through multiple countermeasures, each with limitations and costs. Electronic warfare, useless against the cable guidance, was reoriented toward disrupting Russian reconnaissance drones that identified targets for fiber-optic strikes. This had some success but couldn't eliminate the threat entirely. Small arms fire and portable air defense systems engaged drones when detected, but the low altitude and high speed of attacking drones made intercepts difficult. Specialized counter-drone teams deployed throughout cities, equipped with shotguns, nets, and electronic warfare gear but covering entire urban areas required more teams than could be practically deployed.

Architectural adaptations emerged as residents and soldiers learned to survive in drone-contested environments. Windows were boarded or covered with wire mesh that might snag drones before impact. Interior walls were reinforced with sandbags creating blast-resistant rooms. Movement protocols developed: never linger in open areas, change routes frequently, use covered passageways when possible, and always have escape routes planned. The adaptations reduced vulnerability but couldn't eliminate it, someone had to move supplies, deliver messages, treat wounded, and maintain defensive positions. Each necessary movement created drone attack opportunities.

The fiber-optic penetration demonstrated how technological developments could rapidly negate established tactical advantages. Ukrainian forces had invested heavily in electronic warfare capabilities that effectively defended against radio-controlled drones. Then fiber-optics emerged, rendering those investments largely irrelevant for countering this specific threat. The lesson that technological competition never ended and advantages were always temporary would shape military planning far beyond Ukraine.

The Civilian Toll

The human cost of urban drone warfare manifested in statistics that captured suffering incompletely but undeniably. Ukrainian authorities documented hundreds of civilian casualties from drone strikes in cities within 20 kilometers of front lines during 2024-2025. The real toll was certainly higher, many incidents went undocumented during intense fighting, bodies weren't always recovered, and some casualties were misattributed to artillery or other causes when drones were actually responsible.

The nature of drone casualties differed from conventional artillery or missile strikes. Artillery killed through fragmentation and blast in relatively indiscriminate ways, everyone near an impact point faced similar risk. Drones allowed selective targeting of individuals, often striking single victims or small groups with precision. The result was that some residents of contested cities might live for months without direct threat while neighbors were deliberately targeted and killed. The selectivity made the threat more psychologically destabilizing, you couldn't take comfort in statistical improbability when drones could specifically hunt you.

Particularly disturbing were documented cases of drones deliberately targeting civilians engaged in clearly non-military activities. Russian drones struck people queuing for food at humanitarian distribution points, attacks that served no military purpose beyond spreading terror.

They attacked medical personnel treating wounded, violating laws of war protecting medical workers. They struck children playing in courtyards, elderly residents tending gardens, and families attempting to evacuate. Whether these attacks represented deliberate targeting of civilians or tragic cases of operators mistaking civilians for military targets was debated, but the effect was the same, a civilian population terrorized by constant aerial threat.

The international community's response was vocal but ultimately ineffective. Human rights organizations documented war crimes extensively, building cases for eventual prosecution. Western governments condemned attacks on civilians and provided additional air defense systems. But no amount of condemnation stopped the attacks, and air defense systems designed for faster, higher-flying threats struggled against small, low-altitude drones weaving through urban environments. The gap between international law prohibiting such attacks and practical capability to prevent them illustrated warfare's evolution beyond existing legal and normative frameworks.

Medical infrastructure faced particular challenges. Hospitals and clinics in cities like Kherson operated under constant drone threat. Medical staff couldn't ignore patients when drones approached, they had to continue surgeries, maintain life support, and provide emergency care even while under attack. Several medical facilities suffered direct drone strikes killing patients and staff. The attacks on medical infrastructure, whether deliberate or incidental, violated Geneva Conventions protections but continued nonetheless, demonstrating the conventions' limited practical enforceability during active combat.

Children bore psychological costs that would echo for decades. Growing up in drone-contested cities meant learning to recognize drone sounds before learning to read, understanding that windows were dangerous before understanding why the war was happening, and accepting constant threat as normal childhood experience. Child psychologists working in affected areas reported epidemic rates of anxiety disorders, sleep disturbances, and behavioral problems. An entire generation was being shaped by urban drone warfare in ways that wouldn't fully manifest for years but would influence Ukrainian society long after fighting stopped.

The Landmine-Dropping Drones

Among the most controversial developments in urban drone warfare was Russian employment of quadcopter drones to deliver and scatter anti-personnel landmines in Ukrainian-controlled areas, particularly around Kherson. The tactic combined old and new warfare methods, landmines that had killed and maimed for over a century, delivered by drones that had existed militarily for less than five years in combinations that created nightmarish humanitarian consequences.

The PFM-1 "petal" mine, a Soviet-designed weapon that resembled a flat green plastic shape roughly the size of a hand, was particularly suited to drone delivery. Weighing only 75 grams and designed to be scattered from aircraft or artillery, the mines didn't kill but maimed, typically blowing off feet or causing severe leg injuries when stepped on. Their light weight meant a quadcopter drone could carry dozens, scattering them across parks, streets, residential areas, or anywhere people might walk.

The deliberate use of anti-personnel mines violated the 1997 Ottawa Convention banning such weapons, which 164 nations including Ukraine had signed. Russia, like the United States and China, hadn't ratified the convention and thus wasn't technically bound by its provisions. But the customary international humanitarian law and broader laws of armed conflict still prohibited weapons that caused unnecessary suffering and couldn't distinguish between combatants and civilians. Scattering landmines in civilian areas clearly violated these principles regardless of treaty specifics.

The humanitarian impact was severe and would persist long after combat ended. Unlike conventional munitions that exploded on impact, landmines remained dangerous for years or decades until cleared by specialized teams. Children playing, residents attempting to return home, humanitarian workers delivering aid, all faced risk from mines scattered indiscriminately through urban and suburban areas. The weapons weren't designed to kill but to injure, creating burdens on medical systems and spreading terror through communities where any step might trigger a maiming blast.

Ukrainian authorities documented numerous victims of mine-dropping drone attacks around Kherson and other frontline cities. Farmers working fields, residents collecting firewood, children playing near homes, all suffered amputations from stepping on mines delivered by drones. The casualties weren't just immediate victims but included clearance specialists who risked their lives removing mines, and the broader economic impact as areas became unusable until cleared.

International humanitarian law experts agreed that using drones to scatter anti-personnel mines in civilian areas constituted war crimes requiring prosecution. But as with other drone-related violations, the gap between legal determination and practical enforcement remained vast. Identifying specific individuals responsible for mine-dropping drone missions, from operators flying drones to commanders authorizing tactics to political leaders setting policy presented challenges. Even when responsibility could be established, enforcement required political will and legal mechanisms that didn't yet exist for many drone warfare crimes.

Urban Counter-Drone Tactics

Ukrainian forces developed increasingly sophisticated urban counter-drone tactics through bitter experience and constant innovation. By late 2025, defending against drones in urban environments had become specialized military discipline with dedicated personnel, equipment, and doctrine.

Detection represented the first challenge. Traditional military radar designed for fast-moving aircraft at altitude struggled detecting small, slow-moving drones flying at tree-top level through urban clutter. Ukrainian forces deployed acoustic detection systems that identified drones by their distinctive motor sounds, each drone type having a sonic signature that trained operators could recognize. Visual observers stationed on rooftops provided early warning, though their vulnerability to sniper fire and drone attacks limited effectiveness. Some units experimented with machine learning systems that processed video feeds from security cameras, identifying drones automatically and alerting defenders.

Engagement tactics varied by situation. When drones were detected early at distance, small arms fire, assault rifles, machine guns, even pistols could knock them down. Ukrainian soldiers became surprisingly proficient at leading targets, compensating for drone speed and maneuver. Shotguns proved particularly effective within their limited range, the spread of pellets increasing hit probability against small, maneuverable targets. Some units developed specialized anti-drone shotgun shells with increased range and better patterns optimized for aerial targets.

Nets and physical barriers offered passive defense. Critical locations were surrounded with camouflage netting suspended on poles, creating barriers that drones couldn't penetrate. The netting didn't stop all drones, fiber-optic cables could sometimes be flown through or around nets but complicated attacks enough that many drones crashed or could be engaged while maneuvering around obstacles. Interior defenses included wire mesh over windows and doors, sandbag barriers creating blast-resistant spaces, and reinforced rooms where personnel could shelter during attacks.

Electronic warfare maintained some utility despite fiber-optic guidance. While it couldn't disrupt cable-guided drones, jamming could prevent Russian reconnaissance drones from identifying targets and could disrupt Russian communications coordinating attacks. Signal intelligence helped

Ukrainian forces identify when Russian drone teams were operating nearby, allowing preemptive responses, either counter-battery fire against suspected launch positions or heightened defensive postures anticipating attacks.

Counter-drone drones represented an emerging capability with mixed results. Ukrainian forces developed interceptor drones, typically FPV platforms without warheads but with skilled pilots that could pursue and ram attacking drones, destroying both in mid-air. The tactic required extraordinary pilot skill and favorable conditions but achieved occasional spectacular successes. Videos of interceptor drones chasing and destroying Russian attack drones circulated widely, boosting morale even as the tactic's practical limitations prevented widespread adoption.

Organizational adaptations occurred as units learned to operate in drone-contested environments. Personnel were dispersed more widely, never clustering in groups that presented attractive targets. Command posts moved frequently, never operating from the same location long enough for Russian reconnaissance to identify them. Supply operations occurred at night or used covered routes minimizing exposure. Medical treatment shifted toward distributed first aid rather than centralizing casualties in field hospitals that became drone magnets. Every operational aspect adapted to the constant aerial threat.

The Blurring of Front and Rear

The most profound impact of urban drone warfare was the collapse of distinctions between front-line combat zones and rear areas. Traditionally, military organizations divided territory into forward areas where combat occurred and rear areas where logistics, command, and support functions operated with relative security. The division allowed specialization, combat troops focused on fighting while support personnel focused on sustaining them without constant direct threat.

Urban drone warfare obliterated this comfortable division. A supply depot 15 kilometers behind the "front line" was equally vulnerable to drone strikes as forward positions. Headquarters staff planning operations in city buildings faced the same threats as infantry in trenches. Medical facilities treating wounded operated under fire. The entire depth of urban battlespace became contested requiring everyone, regardless of role or specialization to maintain combat awareness and defensive readiness.

The blurring imposed costs on military efficiency. Support personnel who had to constantly watch for drones couldn't focus entirely on their specialized tasks. Supply operations took longer when every movement required security measures against aerial attack. Command posts operated less efficiently when commanders had to relocate frequently and subordinates couldn't freely travel for face-to-face coordination. The cumulative effect was degradation of military effectiveness even absent successful enemy attacks, the threat alone imposed operational costs.

For civilian populations trapped in contested cities, the blurring meant there was literally no safe place. The apartment building you sheltered in wasn't safe, drones could crash through windows. The street wasn't safe, exposed movement attracted attacks. The basement wasn't safe, drones could be flown into openings. The only safety was distance, evacuating entirely and leaving the urban battlespace. For those who couldn't or wouldn't evacuate, the elderly, the infirm, the stubborn, or those with nowhere else to go life became exercise in constant risk management where death might arrive at any moment from any direction.

The phenomenon also complicated rules of engagement and targeting decisions. When front and rear areas collapsed into single contested space, distinguishing combatants from civilians became more difficult. Someone standing on a street might be a soldier moving between positions or a civilian attempting to reach humanitarian aid. A vehicle might carry military supplies or evacuating families. Buildings might house command posts or refugees. The ambiguity combined with drone operators viewing targets through cameras rather than with their own eyes increased the likelihood of tragic errors where civilians were mistaken for military targets.

Psychological Adaptation and Breaking Points

Human beings demonstrate remarkable capacity for psychological adaptation to adverse circumstances. Residents of drone-contested cities developed coping mechanisms allowing them to function despite constant threat. Some adapted through fatalism, accepting that death might come at any moment and continuing life regardless. Others maintained rigid routines creating the illusion of normalcy. Still others withdrew completely, minimizing exposure by rarely venturing outdoors and conducting only essential activities.

But adaptation had limits, and those limits manifested in various forms of psychological breakdown. Post-traumatic stress disorder became endemic among both military personnel and civilians in heavily contested urban areas. Depression rates soared. Substance abuse increased as people self-medicated to cope with constant anxiety. Domestic violence rose. Suicide rates climbed. The psychological casualties, though often invisible, numbered in the tens of thousands, creating a public health crisis that would burden Ukrainian society for generations.

Children showed particular vulnerability. Child development requires some degree of safety and predictability, foundations shattered by urban drone warfare where threats were constant and unpredictable. Teachers in cities like Kherson reported children unable to concentrate on lessons because they were constantly listening for drones. Parents described children afraid to sleep, refusing to play outside, and developing behavioral problems ranging from extreme clinginess to violent outbursts. The developmental damage inflicted on children growing up in drone-contested cities wouldn't be fully understood for years but would certainly be profound and lasting.

Military personnel faced their own psychological challenges. Soldiers typically understand that combat roles involve danger, they accept risk as part of their function. But support personnel expected relative safety in rear areas. When drones eliminated that safety, psychological impacts were often severe. Supply clerks, headquarters staff, and medical personnel who expected to serve without direct combat faced the same constant threats as infantry, but often with less training or psychological preparation for combat stress. Some adapted, but others broke under pressure they hadn't anticipated when enlisting or accepting assignments.

The aggregate psychological impact on Ukrainian society would require decades to fully manifest and address. An entire population, millions of people experienced or was affected by urban drone warfare in ways that created lasting trauma. The societal cost measured in damaged mental health, disrupted child development, broken families, and diminished human potential would far exceed the direct casualties from drone strikes themselves. Urban warfare had always been psychologically devastating; drones made it ubiquitous and inescapable in ways that previous urban combat, however intense, had not achieved.

Lessons and Implications

The evolution of urban warfare through widespread drone employment taught lessons applicable far beyond the Russia-Ukraine conflict. Military planners worldwide studied Ukrainian experiences, drawing conclusions that would reshape urban warfare doctrine globally.

The first lesson was that future urban warfare would occur in three dimensions, traditional ground-level combat supplemented by constant aerial threat from small drones. Militaries needed capabilities to defend against this threat while employing it against adversaries. Urban warfare training had to incorporate drone tactics and counter-tactics as fundamental skills, not specialized capabilities.

The second lesson was that civilians in contested cities faced unprecedented danger that traditional protections couldn't prevent. International humanitarian law needed updating to address drone warfare realities, though enforcement mechanisms remained problematic. Humanitarian organizations needed new protocols for operating in urban environments where drones could specifically target aid workers and distribution points.

The third lesson was technological, fiber-optic guidance and similar innovations could rapidly overcome electronic warfare defenses, requiring constant innovation in both offensive and defensive capabilities. No defensive system could provide absolute protection, meaning adaptation and resilience were more important than perfect defense.

The fourth lesson concerned psychological impacts, both on military personnel who lost traditional distinction between safe rear areas and dangerous front lines, and on civilians trapped in urban battlespaces where nowhere was truly safe. Future conflicts involving urban drone warfare would generate massive psychological casualties requiring resources and planning equal to physical casualty treatment.

Lieutenant Colonel Viktor Melnyk, who commanded Ukrainian forces defending Kherson through months of intense urban drone warfare, summarized the transformation: "Urban warfare used to be about controlling ground, buildings, streets, and key terrain. Now it's about controlling airspace above the urban environment, which is infinitely more difficult. You can't hold airspace the way you hold a building. You can't fortify against drones the way you fortify against infantry. The entire nature of urban warfare has changed, and we're still learning what that means. What I know is that cities have become more dangerous, more lethal, and harder to defend than ever before. The drone age made urban warfare something fundamentally different from what came before."

The urban warfare evolution driven by widespread drone employment represented one of the most significant tactical developments of the Russia-Ukraine war. The dissolution of front and rear areas, the constant threat to military and civilian alike, the psychological toll of inescapable aerial danger all pointed toward a future where urban combat would be even more devastating than history's bloodiest urban battles. The lessons learned in Ukrainian cities would shape military thinking and urban conflict for decades to come, even as they raised deeply troubling questions about whether modern urban warfare had become so devastating that cities should be evacuated entirely rather than defended or contested. The answer remained unclear, but the question itself represented how profoundly drones had changed warfare's fundamental character.

Chapter 18

The 400-Drone Night

The operations center at Ukraine's Air Force Command fell silent at 10:47 PM on October 9, 2025, as the full scope of what was unfolding became clear. Major Kateryna Levchenko, the night duty officer, stared at her display showing hundreds of radar tracks converging on Ukrainian airspace from multiple directions. She'd managed air defense operations through countless Russian attacks over three years of war, but nothing approached the scale appearing on her screens. The computer was registering over 400 inbound targets, the largest single-night drone assault of the entire conflict.

"Confirm those numbers," she ordered, her voice steady despite the adrenaline flooding her system. "Four hundred-plus?"

"Confirmed," replied the radar operator. "We're tracking 427 distinct contacts as of now, and more launching. They're coming from Crimea, Bryansk, Belarus sectors. Multiple types, Shaheds primarily, but also smaller platforms. This is... I've never seen anything like this."

What unfolded over the next eight hours would test Ukrainian air defenses to their absolute limits, create a diplomatic crisis involving NATO, and demonstrate both the possibilities and the dangers of drone warfare at truly massive scale. The 400-drone night would become a watershed moment, proof that drone saturation attacks could achieve unprecedented scale while simultaneously revealing the vulnerabilities and escalation risks such operations created.

The Attack Begins

Russian planning for the October 9 operation had clearly been extensive. Intelligence assessments would later reveal that Russia had been stockpiling Shahed drones for weeks, accumulating inventory specifically for a massive single-night strike intended to overwhelm Ukrainian defenses and achieve breakthrough damage to critical infrastructure. The operation represented a gamble, expending months of production in one night for maximum shock effect.

The first wave launched at approximately 9:30 PM from Crimean airfields, with roughly 150 Shahed-136 drones departing in staggered formations over thirty minutes. These drones, flying their characteristic low and slow profiles at around 180 kilometers per hour and 50-300 meters altitude, took various routes northward toward Ukrainian cities. Some flew over the Black Sea before turning inland. Others hugged terrain following river valleys and low areas that complicated radar detection. The dispersion forced Ukrainian air defenses to respond across wide frontages rather than concentrating against single approach corridors.

The second wave, launching from Belarus territory around 10:15 PM, added another 120 drones approaching from the north toward Kyiv and other northern cities. This northern attack forced Ukraine to divide already-stretched air defense resources. Fighter aircraft that might have concentrated on the southern threat now had to cover northern sectors as well. Mobile air defense units repositioned to cover likely target areas, but the geographic spread meant some areas would inevitably have thinner coverage.

The third wave was the most concerning, smaller tactical drones launching from Russian positions near the front lines, targeting cities like Kharkiv, Dnipro, and Zaporizhzhia. These 100+ drones, a mix of Shaheds and modified FPV platforms, flew shorter distances but from multiple directions, creating complex three-dimensional air defense problems where threats approached simultaneously from different altitudes, speeds, and vectors.

The final element, approximately 60 drones launching from the Bryansk region in Russia, approached from the east targeting logistics hubs and infrastructure in central Ukraine. The four-axis attack demonstrated sophisticated operational planning, forcing Ukrainian air defenses to respond simultaneously across the entire country rather than concentrating forces against sequential waves.

Major Levchenko's operations center coordinated the defensive response with practiced efficiency despite the unprecedented scale. "Vector all available fighters to the northern sector, establish combat air patrol over Kyiv and Chernihiv," she ordered. "Mobile SAM units to pre-planned positions. All fixed air defense sites weapons free, engage anything confirmed hostile. Get Patriot batteries active, but conserve missiles, we don't know how long this goes."

The Defense Response

Ukrainian air defense had evolved considerably from the war's early days. The integrated air defense system combined Soviet-era equipment with modern Western systems, mobile units with fixed installations, and high-altitude systems with low-level point defenses. Command and control networks linked sensors, weapons, and command posts in near-real-time, allowing coordinated responses that maximized effectiveness while minimizing friendly fire risks.

Fighter aircraft were Ukraine's first line of defense against the Shahed swarms. MiG-29 and Su-27 fighters scrambled from multiple airbases, each aircraft armed with R-27 and R-73 air-to-air missiles plus full cannon loads. The fighters established combat air patrols along likely approach routes, using their onboard radars and ground-based radar cuing to intercept drones before they reached target areas.

Captain Dmytro Shevchenko, flying a MiG-29 callsign "Berkut-3," achieved the first kills of the night at 10:52 PM. "I had multiple contacts on radar, clustered together about fifty kilometers south," he recounted. "Approached from above and behind, got radar lock on the lead contact, fired an R-27. Splash one. Then another. Switched to cannon for the third because I wanted to conserve missiles, we knew this would be a long night. The tracers lit up the darkness as I walked them onto target. Saw the explosion as the Shahed disintegrated. Three down in five minutes, but there were dozens more behind them."

Surface-to-air missile systems engaged across multiple layers. Long-range systems like Patriot and NASAMS engaged drones at maximum detection ranges, often 30-50 kilometers from defended areas. These systems, designed for faster and higher-flying threats, could engage Shaheds but at significant cost, each Patriot missile costing approximately $4 million to destroy a $20,000 drone. Commanders husbanded these expensive missiles, using them only when cheaper systems couldn't engage or when drones threatened the most critical targets.

Medium-range systems provided the backbone of the defense. Soviet-era Buk and S-300 systems, Ukrainian-modified to improve effectiveness against small, slow targets, engaged drones at 15-30 kilometer ranges. These systems, with missiles costing $1-2 million each, still represented unfavorable cost exchanges but were more sustainable than Patriot employment. Ukrainian crews had become remarkably proficient at engaging Shaheds, achieving hit probabilities around 70-80% when conditions were favorable.

Short-range systems and anti-aircraft artillery filled gaps in coverage. Gepard systems, German-provided armored vehicles mounting twin 35mm cannons proved devastatingly effective against slow-moving drones within their 4-kilometer range. Ukrainian crews praised the Gepards' combination of radar fire control and rapid-fire cannons that could engage multiple targets in quick succession. Soviet-era ZSU-23-4 Shilka systems, upgraded with modern fire control, provided similar capabilities. Even towed anti-aircraft guns, technological relics from the Cold War, contributed when positioned along known approach routes.

The most numerous defenders were mobile teams equipped with man-portable air defense systems and small arms. Hundreds of these teams positioned throughout Ukraine engaged drones that penetrated outer defense layers. Teams equipped with Stinger or Igla missiles achieved remarkable success rates, while those with only machine guns and assault rifles still knocked down drones through volume of fire and skill developed over thousands of engagements.

Electronic warfare provided another defensive layer, though with mixed effectiveness. Ukrainian jammers disrupted GPS signals used by some Shaheds, causing navigation failures that sent drones off course or crashing. But many Russian drones incorporated GPS backup systems or inertial navigation allowing continued flight despite jamming. The electronic battle was constant Ukrainian forces jamming, Russian drones adapting, capabilities evolving in real-time technological competition.

The Numbers Game

As the night progressed, the grim arithmetic of the drone war played out across Ukrainian skies. Major Levchenko's operations center tracked shootdowns and penetrations in real-time, updating tallies that determined whether the defense was succeeding or failing.

By midnight roughly two and a half hours into the attack Ukrainian forces had destroyed approximately 220 drones, roughly 52% of the total launched. The success rate was respectable but concerning. At current attrition rates, nearly 200 drones would penetrate defenses to strike targets. The defensive coordination intensified as commanders shifted assets toward sectors where breakthrough seemed imminent.

"Redirect mobile SAM Group Twelve from Kyiv northern approaches to the eastern sector," Levchenko ordered. "We're getting saturation there and losing track of too many contacts. Also, have Patriot Battery Three shift targeting priority to confirmed threats heading toward power infrastructure. We can't let them hit substations."

The interceptor pilots were exhausting their weapons loads and returning to rearm. Captain Shevchenko landed at 12:47 AM having expended all missiles and 80% of his cannon ammunition, claiming seven confirmed shootdowns. "Berkut-3 is Winchester, returning to base for rearm," he'd radioed on approach. "There are still dozens out there. This is insane. I've never seen so many in one night." His aircraft was refueled and rearmed in 23 minutes, a testament to ground crew proficiency developed through years of sustained operations and he was airborne again at 1:17 AM.

The ammunition expenditure was staggering. Patriot batteries fired 47 missiles through the night. NASAMS launched 89. Buk and S-300 systems consumed 234 missiles. Gepard systems fired over 8,000 rounds of 35mm ammunition. Small arms and machine guns expended hundreds of thousands of rounds. The cost in munitions alone exceeded $100 million, sustainable for one night but unsustainable if repeated frequently. Russia was achieving strategic effect even through drones that were shot down, forcing Ukraine to deplete expensive defensive munitions.

The Penetrations

Despite the massive defensive effort, the mathematics of saturation attacks meant significant numbers of drones penetrated to target areas. Ukrainian authorities later confirmed that approximately 30 drones, roughly 7% of the total launched successfully struck targets, a testament to defensive effectiveness but still representing substantial damage.

The confirmed hits targeted Ukraine's power infrastructure with precision that suggested detailed intelligence preparation. A Shahed struck the Zmiiv Thermal Power Plant near Kharkiv, causing a fire that forced one generating unit offline for three weeks. Another hit a substation outside

Dnipro, creating cascading blackouts affecting 300,000 residents. A third struck fuel storage facilities in Lviv, destroying 2,000 tons of diesel fuel in spectacular fires visible across the city.

Kyiv, the primary target of the northern wave, suffered multiple impacts despite concentrated defenses. Drones struck a heating plant in the Holosiivskyi district, a transformer station in Darnytskyi, and a water pumping facility that temporarily disrupted service to 150,000 apartments. The attacks were clearly intended to inflict maximum disruption as autumn weather turned cold and reliable utilities became critical for civilian survival.

Other penetrating drones hit less critical but still significant targets. A warehouse complex in Odesa, a railway repair facility in Poltava, and a fuel depot in Chernihiv all suffered damage. Some drones apparently missed their intended targets or suffered navigation failures, one crashed into a field 50 kilometers from any obvious target, another into a forest, and a third into a reservoir. These failures demonstrated that even sophisticated attack planning couldn't guarantee success against defended targets over long ranges.

The most tragic incident occurred in Kharkiv, where a Shahed apparently off course or suffering navigation failure struck an apartment building, killing seven civilians including two children. The strike demonstrated the indiscriminate danger that saturation drone attacks posed to civilian populations, even drones aimed at military or infrastructure targets could kill civilians through malfunction, defensive evasion, or terminal guidance failure.

The Polish Intrusion

The most diplomatically significant development of the 400-drone night occurred in the northwestern quadrant, where Ukrainian territory bordered Poland, a NATO member operating under Article 5 collective defense commitments. As Ukrainian air defenses engaged the northern attack wave, a subset of drones veered westward, either through navigation errors, defensive evasion, or possibly deliberate Russian testing of NATO responses.

At 11:34 PM, Polish air defense radar detected multiple airborne contacts crossing from Ukrainian into Polish airspace near the border city of Przemyśl. Initial assessment identified 19 distinct contacts, all consistent with Shahed-136 profiles, flying at low altitude and slow speed. The contacts were clearly not military aircraft, their radar signatures, flight profiles, and speeds matched unmanned systems. But their presence in Polish airspace constituted a violation of NATO territory during an active conflict.

The Polish Air Force implemented pre-planned response protocols. F-16 fighters from the 31st Tactical Air Base at Poznań scrambled within minutes, racing southeast toward the border area. Ground-based air defense systems went to highest alert status. Polish authorities notified NATO headquarters, triggering alliance consultations about potential collective defense implications. The situation was unprecedented, Russian weapons systems, albeit unmanned, had entered NATO territory in substantial numbers during active hostilities.

Polish fighters established visual contact with the intruding drones at 11:51 PM. The F-16s, equipped with targeting pods providing high-resolution video, confirmed the drones were Shaheds identical to those attacking Ukraine. "We could see them clearly," reported Major Krzysztof Nowak, leading the Polish intercept. "Flying low, following terrain, definitely not Ukrainian. Whether they crossed intentionally or by error didn't matter; they were in Polish airspace and they were weapons."

The rules of engagement were clear but politically fraught. Poland had legal authority to destroy anything in Polish airspace without permission. But doing so meant NATO weapons actively engaging Russian weapons, even unmanned ones, creating escalation risks that extended beyond Poland to the entire alliance. NATO Secretary General, awakened for emergency consultations, authorized Polish engagement under self-defense provisions but urged restraint and careful documentation to support diplomatic protests.

Polish fighters engaged at 12:07 AM, destroying four Shaheds with cannon fire over unpopulated areas. The remaining fifteen drones, possibly detecting the fighters or responding to pre-programmed routing, turned back eastward and reentered Ukrainian airspace before additional engagements occurred. Ukrainian air defenses destroyed twelve of these drones; three apparently crashed or were unaccounted for.

The Diplomatic Firestorm

The Polish airspace incursion created an immediate diplomatic crisis that overshadowed even the massive attack on Ukraine. NATO ambassadors convened emergency consultations at 2:00 AM Brussels time, with alliance members debating appropriate responses to what some characterized as Russian attack on NATO territory.

Eastern European NATO members, Poland, Lithuania, Latvia and Estonia advocated strong responses including direct military aid to Ukraine, potentially including weapons systems previously restricted, and stern warnings to Russia about consequences of further airspace violations. These nations, acutely aware of their geographic proximity to Russia and historical experience with Russian aggression, viewed the incursion as testing NATO resolve and requiring firm pushback.

Western European members of Germany, France and Italy counseled caution, emphasizing that the incursion involved unmanned systems that might have strayed unintentionally, that no casualties or damage occurred on NATO territory, and that measured diplomatic responses would be more appropriate than military escalation. These nations, generally more cautious about confrontation with nuclear-armed Russia, worried that overreaction could trigger an escalation spiral leading to direct NATO-Russia conflict.

The United States, as alliance leader, occupied middle ground. American officials condemned the airspace violation while acknowledging uncertainty about whether the drones deliberately targeted Poland or strayed due to navigation failures or defensive evasion. The U.S. supported Poland's right to defend its airspace while urging alliance unity and avoiding responses that might provide Russian pretexts for further escalation. The American position reflected competing pressures, supporting NATO allies while avoiding direct conflict with Russia that might escalate to nuclear dimensions.

Russia's response was characteristically duplicitous. Official Russian statements claimed complete ignorance of any airspace violations, suggesting either Ukrainian drones or Ukrainian disinformation were responsible. When confronted with Polish radar data and video footage clearly showing Russian Shaheds in Polish airspace, Russian officials shifted to claiming the drones had malfunctioned or been driven off course by Ukrainian air defense actions, essentially blaming Ukraine for Russian weapons entering NATO territory. The responses convinced nobody but demonstrated Russian unwillingness to acknowledge actions that might trigger formal NATO responses.

The diplomatic outcome was ultimately restrained. NATO issued a strongly-worded statement condemning the airspace violation and warning Russia that future incursions would face "appropriate responses" without specifying what those might be. Poland received verbal support from all alliance members and pledged additional air defense systems to prevent future incidents. Behind closed doors, NATO conveyed to Russia through back channels that further airspace violations might trigger Article 4 consultations, the step before invoking Article 5 collective defense making clear that patience had limits.

Strategic and Tactical Lessons

The 400-drone night, viewed holistically after the immediate chaos subsided, provided crucial insights into drone warfare at massive scale. Ukrainian military analysts, NATO observers, and independent researchers all studied the attack extensively, drawing lessons that would influence future operations and planning.

The most obvious lesson concerned scale: Russia could generate and employ truly massive drone swarms when willing to expend accumulated inventory. The 400-plus drone attack represented probably 8-10 weeks of Russian production launched in a single night. The tactic achieved psychological and operational shock effects while forcing Ukraine to expend enormous defensive resources. Whether the exchange favored Russia strategically remained debatable, destroying infrastructure worth billions while expending drones worth perhaps $10-15 million suggested favorable return on investment, but only if Russia could sustain such attacks repeatedly, which production rates didn't support.

The defensive success rate of approximately 93% of attacking drones destroyed demonstrated Ukrainian air defense effectiveness while highlighting the remaining vulnerability. Even 7% penetration could cause significant damage when the attack involved hundreds of drones. Perfect defense proved impossible; Ukraine needed offensive capabilities to strike Russian drone production and launch facilities, reducing attack volume at source rather than relying entirely on defensive interception.

The ammunition expenditure crisis became undeniable. Ukraine fired hundreds of millions of dollars in defensive munitions to counter tens of millions in attacking drones. The economics were unsustainable without continued Western support, and even Western stockpiles weren't infinite. Future air defense required cheaper interception methods, whether through electronic warfare improvements, directed energy weapons, or counter-drone drones that could engage threats without prohibitive ammunition costs.

The Polish incursion demonstrated escalation risks inherent in large-scale drone operations. Navigation errors, defensive evasion, or deliberate provocation could send drones into neighboring countries' airspace, creating international incidents with unpredictable consequences. The incident was contained through restraint and established alliance mechanisms, but luck played a role, had drones struck Polish territory causing casualties, political pressure for NATO response might have been irresistible regardless of strategic rationality.

The coordination requirements for defending against distributed, multi-axis attacks became clear. Ukrainian air defense succeeded because of integrated command and control allowing flexible response to rapidly changing threat pictures. Militaries lacking such integration would struggle against similar attacks. The lesson extended beyond Ukraine, any nation facing potential drone threats needed investment in air defense integration, not just hardware acquisition.

The Psychological Impact

Beyond tactical and strategic lessons, the 400-drone night left psychological imprints on everyone involved. For Ukrainian civilians, the attack reinforced awareness that nowhere was truly safe, that Russian capability to strike deep into Ukraine remained potent despite three years of war. The massive scale generated fear that exceeded actual damage, people imagined worse scenarios than occurred, extrapolating from 400 drones to potential future attacks of 600, 800, or more.

Ukrainian military personnel experienced different psychological effects. Air defense forces felt pride in achieving 93% shootdown rate against unprecedented attack volume, validation of years spent developing tactics, training, and coordination. But they also felt exhaustion and concern about sustainability. "We performed brilliantly," acknowledged one Patriot battery commander. "But we also got lucky if Russia had launched 500 or 600 drones, or if they'd concentrated against fewer targets, penetration rates would have been higher. We're good, but we're not invincible. And we're burning through munitions faster than they can be replaced."

International observers drew sobering conclusions about future warfare. If a nation of Russia's modest military-industrial capacity could generate 400-drone attacks, what might more capable nations achieve? China, with far larger manufacturing base and technological sophistication, could presumably generate drone swarms of thousands if it chose to invest resources. The United States, Israel, or other advanced military powers could likely exceed even those numbers. The 400-drone night provided a glimpse of future warfare that many found deeply unsettling.

Russian reactions were mixed. State media portrayed the attack as a demonstration of Russian military power, showcasing the ability to strike Ukrainian infrastructure despite Western aid. Military bloggers and analysts offered more nuanced assessments, noting the high attrition rates and questioning whether expending months of production in a single night represented a wise strategy. Some recognized that Ukraine's defensive success suggested Russian drone attacks, however large, couldn't achieve decisive strategic effects without complementary ground offensives that Russia currently couldn't generate.

Long-term Implications

The 400-drone night of October 9, 2025, would be remembered as a pivotal moment in the drone war's evolution. It demonstrated both the potential and limitations of mass drone warfare, provided crucial data about defense effectiveness and vulnerabilities, and created diplomatic precedents that would influence future conflicts involving NATO borders and drone warfare.

For Ukraine, the attack validated continued investment in integrated air defense while highlighting the need for offensive capabilities targeting Russian drone production. The 93% shootdown rate proved defenses worked but weren't sufficient. Ukraine needed to attack the problem at its source by striking launch sites, storage facilities, and manufacturing plants deep inside Russia.

For Russia, the attack demonstrated capability to generate massive strikes but uncertain strategic value. The resources expended both drones and the missiles launched simultaneously didn't produce breakthrough damage justifying the cost. Future Russian strategy might emphasize sustained smaller attacks over rare massive strikes, though the psychological value of spectacular large-scale operations couldn't be discounted.

For NATO, the Polish airspace incursion forced serious consideration of response protocols for future incidents. The alliance needed clearer guidelines about responding to unmanned systems entering NATO territory, guidelines balancing member nations' right to self-defense against escalation risks from overreaction. The discussion would continue through subsequent months, eventually producing updated rules of engagement and response protocols applicable throughout the alliance.

For military planners worldwide, the 400-drone night became a case study in drone warfare at scale. It demonstrated that determined adversaries with modest resources could generate attacks of hundreds of drones, that defenses could achieve high success rates but not perfect protection, and that such operations created escalation risks through territorial spillover. The lessons would shape military procurement, doctrine development, and strategic planning for decades.

Major Kateryna Levchenko, who had coordinated Ukrainian air defense through that historic night, reflected on the experience: "When I saw 400 drones on my screen, I thought 'this is impossible, we can't stop this.' But we did, we stopped 93% of them through skill, coordination, and determination. That night proved what we're capable of. It also proved we face an enemy willing to expend enormous resources attacking us, and that this war is far from over. We won that night's battle. The war continues."

The sun rose on October 10, 2025, over a Ukraine scarred by new strikes but still standing, its air defenses proven resilient against even massive drone swarms, its people exhausted but unbroken. The 400-drone night entered history as both demonstration of drone warfare's terrible potential and testament to human capacity to defend against it. The equilibrium between offense and defense, between attack and protection, remained precarious but for that night at least, defense had prevailed.

Chapter 19

Types and Platforms

By the autumn of 2025, the skies over Ukraine had become a complex ecosystem of unmanned aerial vehicles, each type filling a specific tactical niche that had been discovered, refined, and perfected through brutal trial and error. What began in February 2022 with civilian DJI Mavics pressed into military service had evolved into a sophisticated taxonomy of purpose-built platforms, each optimized for particular missions along the vast front lines and deep into enemy territory.

The drone war had matured beyond improvisation. Both Ukraine and Russia now fielded diverse fleets of unmanned systems, ranging from tiny racing quadcopters costing less than $500 to sophisticated long-range strike platforms carrying payloads across hundreds of kilometers. Understanding these platforms, their capabilities, limitations, and tactical employment had become essential to comprehending modern warfare itself.

FPV Kamikaze Drones: The Workhorses

The First-Person View kamikaze drone emerged as the defining weapon system of the conflict, accounting for the vast majority of drone strikes on both sides. These were the descendants of racing drones, adapted for war with deadly efficiency. A typical Ukrainian or Russian FPV drone in 2025 measured between seven and thirteen inches diagonally across its motor-to-motor span, weighed two to four kilograms fully loaded, and cost between $400 and $800 to manufacture.

The basic configuration had become standardized through thousands of iterations in combat. Four brushless motors drove carbon fiber propellers at speeds exceeding 100 kilometers per hour. A lithium polymer battery provided ten to fifteen minutes of flight time, though combat missions rarely lasted more than seven minutes from launch to impact. The warhead, typically a reshaped anti-tank grenade, RPG warhead, or purpose-molded explosive charge nested in the drone's center, adding one to three kilograms to the airframe.

What made these drones revolutionary was the control system. Unlike traditional military UAVs controlled via satellite links or line-of-sight radio, FPV drones transmitted real-time video directly to an operator wearing goggles, creating an immersive cockpit view that allowed pilots to thread through forests, dive into trenches, and maneuver around obstacles with uncanny precision. The pilot flew the drone as if sitting in a tiny cockpit, seeing exactly what the camera saw, experiencing the flight with an intimacy that dramatically improved accuracy.

The operational range of radio-controlled FPV drones reached five to eight kilometers under ideal conditions, though electronic warfare typically compressed this to two or three kilometers on active fronts. Pilots launched from concealed positions, often from reinforced dugouts or camouflaged vehicles, and guided their charges toward targets identified by reconnaissance drones or ground observers. The attack profile varied by target: diving attacks on armored vehicles to strike weak top armor, horizontal approaches for bunkers and trenches, or hovering searches for personnel in tree lines.

By October 2025, both sides were losing hundreds of FPV drones daily consumed in successful strikes, shot down by enemy fire, crashed due to jamming, or simply lost to battery depletion. The attrition rate was extraordinary, but so was the production capacity. Ukraine alone was manufacturing over 125,000 FPV drones monthly, feeding an insatiable appetite for these expendable precision weapons.

The evolution toward larger platforms, the shift from seven-inch to ten-inch and finally thirteen-inch frames reflected lessons learned about payload capacity and wind resistance. Larger drones carried bigger warheads and additional equipment like backup radio systems or fiber-optic cable spools. They also performed better in the variable weather conditions of Ukrainian seasons, maintaining control in winds that would ground their smaller predecessors.

Strike Quadcopter Bombers

While FPV drones were one-way weapons, strike quadcopters offered reusability and flexibility. These platforms, descended from commercial photography drones, loitered over target areas and dropped munitions with surprising precision. The typical strike quadcopter resembled a scaled-up DJI Mavic or Matrice, with four or six rotors arranged around a central body housing the battery, flight controller, and camera systems.

Ukrainian forces favored domestically-produced variants that solved the supply chain problems created when China restricted DJI drone exports. Models from manufacturers like Escadrone and TAF Drones carried payloads of three to eight kilograms and could remain airborne for twenty to forty minutes depending on load. The control range extended to ten kilometers or more using enhanced radio systems, and some advanced models incorporated satellite communication for beyond-line-of-sight operations.

The munitions themselves evolved into a dark art. Early in the war, soldiers had simply taped grenades to commercial drones, creating crude but effective bombing platforms. By 2025, purpose-designed bomblets filled the cargo bays. Some were modified mortar shells with 3D-printed fins for stability. Others were purpose-molded explosives with shaped charges for anti-armor work. The most controversial were the small anti-personnel mines, internationally prohibited weapons that Russian quadcopters had been documented dropping across occupied territories and particularly in places like Kherson.

The release mechanisms ranged from simple servo-actuated claws to sophisticated multi-station racks that allowed a single drone to engage several targets per sortie. Advanced strike quadcopters used laser rangefinders or visual AI to calculate precise release points, accounting for altitude, wind, and target movement. A skilled operator could place a bomblet into a trench from 100 meters altitude with meter-level accuracy.

The tactical employment differed markedly from FPV operations. Strike quadcopters hunted at dawn and dusk, using reduced visibility to approach targets. They worked in pairs or trios, with one drone designating targets while others attacked. They supplemented artillery, hitting targets too small or too fleeting for conventional fires. And crucially, they returned to base, allowing expensive sensors and flight systems to be used again and again.

Russian forces employed similar platforms, with particular emphasis on night operations using thermal imaging. Their strikes on Ukrainian rear areas, reaching dozens of kilometers behind the front lines, demonstrated how these reusable platforms extended battlefield depth. The psychological impact of knowing that a quadcopter might be hovering silently overhead, invisible in darkness, its thermal camera scanning for signs of life, was considerable.

Reconnaissance Winged Drones

While quadcopters dominated close combat, fixed-wing reconnaissance drones owned the operational depth. These platforms, with wingspans of two to four meters, launched from catapults or runways and flew patterns at altitudes of 1,000 to 4,000 meters, well above small arms range and often above the ceiling of portable air defense systems.

Ukrainian reconnaissance drones like variants of the Leleka-100 and domestically-produced designs provided commanders with real-time intelligence over areas measuring tens of square kilometers. Equipped with electro-optical and infrared cameras, these platforms could detect vehicle movements, troop concentrations, and defensive preparations from safe distances. Flight endurance ranged from two to six hours, allowing extended surveillance of key areas.

The missions flown by reconnaissance drones were unglamorous but essential. They conducted pre-strike reconnaissance, identifying targets for artillery or FPV teams. They performed battle damage assessment after attacks. They monitored enemy supply routes and identified logistics hubs. They tracked enemy movements during offensive operations. Every successful strike, every ambush avoided, every well-placed artillery barrage began with reconnaissance drone footage.

The evolution of these platforms focused on survivability and sensor quality. Later models incorporated radar warning receivers to detect air defense systems. Some used reduced radar cross-sections and quiet electric motors to avoid detection. The sensor packages improved dramatically, with high-resolution cameras capable of reading vehicle identification numbers from 2,000 meters and thermal imagers that could detect the heat signature of a truck engine from five kilometers.

Both sides devoted substantial effort to shooting down enemy reconnaissance drones. These platforms, worth thousands of dollars and carrying sensitive equipment, became priority targets. Mobile air defense units hunted them with missiles and autocannon. Electronic warfare systems jammed their control links, sending them tumbling from the sky or forcing autonomous return-to-base protocols. Fighter aircraft, when available, intercepted them with missiles or gunfire.

The cat-and-mouse game between reconnaissance drones and air defenses drove a cycle of innovation. Drones flew higher to avoid short-range air defense, so defenses deployed longer-range missiles. Drones adopted evasive flight patterns and operated at night, so defenses added radar and infrared sensors. Some reconnaissance drones began carrying defensive payloads, chaff dispensers or infrared flares borrowing concepts from manned aviation.

Long-Range Deep-Strike Platforms

The strategic dimension of the drone war manifested in long-range strike platforms capable of attacking targets hundreds of kilometers from launch points. These weapons blurred the line between drone and cruise missile, combining the loitering capability and retargetability of UAVs with the range and payload of traditional missiles.

Ukraine's domestically-developed platforms like the Queen Hornet represented the pinnacle of this category. With ranges exceeding 1,000 kilometers, these drones could strike Russian oil refineries in Tatarstan, logistics hubs in Rostov, and airfields in Crimea from launch sites deep inside Ukrainian-controlled territory. The typical deep-strike drone measured three to five meters in length with a wingspan of two to four meters, powered by a small piston engine or electric motor depending on range requirements.

The payload capacity of these platforms ranged from 20 to 75 kilograms of high explosive, sufficient to damage industrial facilities or destroy parked aircraft but modest compared to traditional cruise missiles. What they lacked in destructive power, they compensated for in numbers and cost. A Ukrainian long-range strike drone cost perhaps $50,000 to $100,000 to produce, a fraction of the price of a cruise missile allowing mass production and mass employment.

Navigation systems combined GPS, inertial guidance, and terrain-matching algorithms. The drones flew low to avoid radar detection, following pre-programmed routes that exploited gaps in air defense coverage. Some variants incorporated optical or infrared seekers for terminal guidance, allowing them to identify and strike specific buildings or vehicles within larger complexes.

Russia fielded the infamous Shahed-136, rebranded as the Geran-2, as its primary long-range strike weapon. Originally Iranian-designed and manufactured, by 2025 Russia was producing these platforms domestically using smuggled Western electronics and Chinese components. The Shahed's distinctive delta wing and pusher propeller configuration became synonymous with the terror bombing campaign against Ukrainian cities and infrastructure.

The Shahed measured 3.5 meters in length with a 2.5-meter wingspan, constructed primarily from foam and plywood to reduce costs and radar signature. Its 50-kilogram warhead and 1,000-kilometer range made it ideal for attacking power stations, fuel depots, and command centers. The distinctive whining buzz of its piston engine earned it the nickname "moped" among Ukrainians, a sound that became the soundtrack to countless sleepless nights.

Both sides employed these long-range platforms in coordinated packages. A typical Russian strike might combine Shahed drones with cruise missiles and ballistic missiles, overwhelming Ukrainian air defenses through sheer variety. The drones, being slower and lower-flying, absorbed interceptor missiles and drew fire while faster missiles slipped through. Ukrainian deep-strike operations focused on economic targets, the oil refineries and storage facilities that fueled Russia's war machine achieving strategic effects through cumulative damage.

Interceptor Drones: Counter-Drone Missions

As the skies filled with hostile UAVs, both sides experimented with dedicated counter-drone platforms. These interceptor drones ranged from modified FPV quadcopters to purpose-built designs optimized for air-to-air engagement. The concept was simple: meet enemy drones with friendly drones, fighting fire with fire.

Ukrainian forces developed FPV interceptors specifically to engage Russian reconnaissance drones and incoming Shaheds. These platforms carried no explosive payload, instead relying on high-speed collision to destroy their targets. The pilots, using the same FPV control systems as attack drone operators, would identify enemy aircraft and execute ramming attacks, sacrificing their drone to bring down a more valuable enemy asset.

The mathematics were compelling. A $500 FPV drone could potentially destroy a $20,000 reconnaissance platform or a $50,000 Shahed, creating enormous cost asymmetry. Success rates varied, but experienced interceptor pilots achieved kills in 20-30% of attempts, respectable odds given the relative costs.

Other counter-drone approaches included net-carrying quadcopters designed to entangle enemy drones, and platforms equipped with signal jammers to disrupt enemy control links at close range. Some experimental systems mounted directed energy weapons, essentially powerful lasers to burn out enemy sensors or flight control systems.

The tactical employment of interceptor drones required different skills than attack operations. Pilots needed to anticipate enemy drone behavior, predict flight paths, and execute precise intercepts at combined closing speeds exceeding 150 kilometers per hour. The best interceptor pilots came from backgrounds in first-person view racing, where split-second reflexes and spatial awareness determined success.

Despite promising results in specific engagements, interceptor drones remained a niche capability through 2025. The sheer volume of enemy drones often overwhelmed the available interceptors, and traditional air defense systems, autocannon, surface-to-air missiles, and electronic warfare continued to bear the primary burden of counter-drone operations. Nevertheless, the concept proved viable and pointed toward future developments in autonomous air-to-air combat.

Multi-Rotor vs. Fixed-Wing: Tactical Niches

The fundamental division between multi-rotor (quadcopter and hexacopter) and fixed-wing platforms reflected different tactical requirements and operational constraints. Each configuration offered distinct advantages that commanders learned to exploit.

Multi-rotor drones excelled at precise maneuvering in confined spaces. They could hover, fly sideways or backwards, and execute sharp direction changes impossible for fixed-wing aircraft. This made them ideal for urban combat, forest operations, and attacks on fortified positions where final approach required threading between obstacles. The vertical takeoff and landing capability eliminated the need for runways or catapults, allowing deployment from anywhere, vehicles, buildings, trenches, or forest clearings.

The limitations were equally clear: poor endurance, limited range, and vulnerability to wind. Multi-rotors consumed tremendous power to remain airborne, restricting flight times to 15-40 minutes depending on size and payload. They struggled in wind speeds above 30 kilometers per hour, which grounded operations during storms or reduced accuracy during breezy conditions. The relatively slow forward speed, typically 50-80 kilometers per hour for strike quadcopters made them vulnerable to interception.

Fixed-wing platforms traded maneuverability for endurance and range. A reconnaissance drone could loiter for hours, and long-range strike platforms could reach targets hundreds of kilometers away. The aerodynamic efficiency of wings provided far greater range per unit of battery capacity or fuel. Fixed-wing drones also flew faster, typically 80-150 kilometers per hour making them harder to intercept and more survivable in contested airspace.

The disadvantages included launch and recovery requirements, larger physical size, and inability to maneuver in confined spaces. Fixed-wing drones needed catapults or runways for takeoff, and parachutes or nets for recovery. They couldn't hover or reverse direction, limiting their utility for close-range tactical missions.

Commanders learned to employ both types in complementary roles. Fixed-wing reconnaissance drones provided the operational picture, identifying targets and tracking enemy movements. Multi-rotor strike platforms then attacked those targets with precision. FPV kamikaze drones, themselves multi-rotors, provided the tactical firepower for immediate, responsive strikes. Long-range fixed-wing platforms conducted strategic strikes on distant targets.

The mix varied by mission and terrain. In the urban combat of cities like Bakhmut or Mariupol, multi-rotors dominated. In the open terrain of southern Ukraine, fixed-wing platforms proved more useful for reconnaissance. Both types flooded the battlefield in numbers that would have seemed impossible in 2022.

Ground Control Stations and C2 Infrastructure

Behind every drone flying combat missions stood a ground control station and the command and control infrastructure that enabled coordinated operations. By 2025, these systems had evolved from improvised laptops and hand-built radio sets into sophisticated mobile command posts.

A typical Ukrainian FPV drone team operated from a concealed position within a few kilometers of the front line. The control station consisted of a high-gain directional antenna, a video receiver tuned to the drone's transmission frequency, and a radio controller for the pilot. The pilot wore FPV goggles, essentially small screens mounted inches from the eyes displaying the real-time video feed from the drone's camera. With practice, the immersion was complete; pilots reported feeling as though they were actually flying, experiencing vertigo during aggressive maneuvers.

More sophisticated control stations for reconnaissance and strike quadcopters used ruggedized tablets or laptop computers running custom ground control software. These interfaces displayed telemetry, altitude, speed, battery voltage, GPS position alongside the video feed. Multiple windows allowed operators to monitor several drones simultaneously or coordinate with artillery fire control centers.

The command and control challenge scaled with the number of drones. A single infantry company might operate a dozen FPV drones, three or four strike quadcopters, and share feeds from battalion-level reconnaissance platforms. Coordinating these assets while avoiding frequency conflicts, managing limited numbers of trained pilots, and ensuring tactical integration with conventional forces required new organizational structures.

Both sides developed specialized drone units with dedicated chains of command. Russia's Rubicon elite drone unit exemplified this approach, concentrating expert pilots and advanced equipment under specialized leadership. Ukrainian forces often embedded drone teams directly with infantry companies, ensuring tight integration between aerial fires and ground maneuvers.

The communications infrastructure supporting drone operations became a critical vulnerability and a priority target. Both sides attacked enemy control stations with artillery, FPV strikes, and even special forces raids. The loss of a control station not only eliminated the ground equipment but also effectively neutralized all drones that the team could operate.

Hardening these facilities against attack drove innovations in mobility and concealment. Control stations moved frequently, operating from reinforced vehicles, underground facilities, or rapidly-emplaced camouflaged positions. The radio signature of drone operations, both the control signals and the video downlinks made concealment challenging, and both sides developed direction-finding equipment to locate enemy control stations for targeting.

The Integration Challenge

By October 2025, the sheer variety of drone types operating simultaneously over Ukrainian battlefields created complex integration challenges. A commander planning an offensive might coordinate reconnaissance wings surveying enemy positions, FPV strikes softening defensive points, strike quadcopters attacking artillery pieces, and long-range platforms hitting logistics hubs, all while friendly air defense systems attempted to shoot down enemy drones performing similar missions.

The solution emerged through a combination of technology and doctrine. Deconfliction zones separated friendly drones by altitude, with close-range FPV operations below 100 meters, strike quadcopters at 100-300 meters, and reconnaissance wings above 1,000 meters. Temporal deconfliction assigned different units exclusive time windows for operations in contested airspace. Geographic deconfliction divided the battlefield into sectors with designated responsible units.

Digital coordination systems allowed real-time awareness of friendly drone positions. Transponders on Ukrainian drones transmitted identification signals to prevent friendly fire from air defense systems. Shared digital maps displayed drone operations, allowing artillery to avoid firing while friendly drones transited their guns' trajectories, and vice versa.

Despite these measures, integration remained imperfect. Friendly fire incidents destroyed numerous drones throughout 2025. The fast tempo of combat operations sometimes prevented proper coordination, with units launching strikes without informing adjacent units. The combination of conventional air defense, counter-drone systems, and thousands of drones from both sides created an extraordinarily complex aerial environment.

Yet somehow it worked. The learning curve was steep and casualties among drones were enormous, but both Ukrainian and Russian forces managed to operate diverse fleets of UAVs with increasing sophistication. The platforms themselves, FPV kamikazes, strike quadcopters, reconnaissance wings, long-range strikers, and experimental interceptors each found their place in the new paradigm of warfare.

These were the tools that transformed the Russia-Ukraine conflict into history's first true drone war. Understanding them, their capabilities and limitations, their costs and their casualties was essential to understanding the war itself. And as these platforms proliferated beyond Ukraine's borders, understanding them became essential to understanding the future of warfare worldwide.

Chapter 20

Manufacturing Ecosystem

The transformation of Ukraine from drone importer to drone superpower in less than three years stands as one of the most remarkable industrial achievements of modern warfare. By October 2025, Ukraine's drone manufacturing ecosystem had become a sprawling network of factories, workshops, volunteer groups, and combat units, a decentralized industrial complex producing over 200,000 drones monthly and employing tens of thousands of workers. This chapter explores how Ukraine built this capability from almost nothing, the challenges of wartime production, and the organizations that made it possible.

TAF Drones: The Giant

When discussing Ukrainian drone manufacturing at scale, all roads lead to TAF Drones. By 2025, TAF had emerged as the largest single manufacturer in Ukraine's defense industrial base, a company whose output alone exceeded the entire military production of many NATO countries. The numbers were staggering: 40,000 drones per month rolling off production lines, valued at over $1 billion annually. To put this in perspective, TAF was producing more drones in a single month than most countries' entire military inventories.

TAF's facility operated around the clock in shifts, a hive of activity that never stopped. The company had evolved far beyond its origins, scaling up from small-batch production to true mass manufacturing with assembly lines that would have seemed impossible in early 2022. The workforce numbered in the thousands, technicians, assemblers, quality control specialists, programmers, and logistics coordinators working in synchronized choreography to maintain the relentless pace of production.

The manufacturing floor at TAF embodied the industrialization of drone warfare. Unlike the improvised workshops of 2022, this was a modern factory with specialized work stations, automated testing equipment, and quality control protocols borrowed from automotive manufacturing. Yet it retained the urgency of wartime production, no bureaucratic delays, no lengthy approval processes, just the constant pressure to produce more drones faster.

The product line diversified across multiple platforms. TAF manufactured FPV kamikaze drones in several size categories, seven-inch, ten-inch, and thirteen-inch frames optimized for different payloads and ranges. The company produced strike quadcopters for bomber missions, reconnaissance platforms for intelligence gathering, and specialized variants with fiber-optic guidance systems. Each product line had its own assembly process, its own quality standards, its own supply chain.

Supply chain management became TAF's competitive advantage. While smaller manufacturers struggled with component shortages, TAF's scale allowed bulk purchasing and strategic stockpiling. The company maintained relationships with dozens of suppliers across Ukraine and abroad, ensuring redundancy in critical components. When one supplier faltered, alternatives stood ready. When specific electronic components became scarce, engineering teams redesigned circuit boards to use available substitutes.

The company's vertical integration increased throughout 2024 and 2025. TAF began manufacturing components that had previously been sourced externally, carbon fiber frames molded in-house, circuit boards assembled from raw components, even electric motors wound in company facilities. This vertical integration served multiple purposes: it reduced dependence on external suppliers, lowered per-unit costs through economies of scale, and allowed tighter quality control.

Quality assurance at TAF's scale required systematic approaches. Every drone underwent flight testing before shipment, with technicians piloting test flights in secure facilities to verify motor function, video transmission, and control responsiveness. Electronic components passed through automated testing rigs that checked voltage, current draw, and signal quality. The rejection rate hovered around 3-5%, low by wartime standards but representing thousands of drones monthly that failed to meet specifications.

The logistics of distributing 40,000 drones monthly across Ukrainian military units presented its own challenges. TAF operated a network of regional distribution centers, each stocked with drones and spare parts. Military units placed orders through a digital system, and deliveries arrived via military transport within days. The system worked remarkably well, though combat losses sometimes exceeded delivery rates during intensive operations.

TAF's workforce represented a cross-section of Ukrainian society. Many employees were women, taking on roles in assembly, quality control, and logistics while male relatives served at the front. Some workers were veterans themselves, invalided out of combat but contributing to the war effort through manufacturing. Others were civilians who had fled occupied territories, finding employment and purpose in drone production. The wages were competitive by Ukrainian wartime standards, and the knowledge that their work directly supported soldiers motivated performance.

The company's innovation pipeline never stopped. Engineering teams constantly refined designs based on combat feedback. When operators reported control issues at long range, engineers redesigned antenna systems. When jamming increased, new frequency-hopping protocols were developed. When fiber-optic drones emerged, TAF reverse-engineered the concept and began mass production within months. The cycle time from battlefield problem to engineering solution to mass production measured in weeks, not years.

By late 2025, TAF was beginning to look beyond Ukraine's borders. Company executives spoke openly about post-war export potential, positioning Ukraine as a global drone manufacturing hub. NATO countries were already studying Ukrainian designs and operational concepts, and some had expressed interest in purchasing Ukrainian-made systems. TAF's wartime experience, building reliable, combat-proven platforms at scale and low cost positioned it uniquely in the global defense market.

Wild Hornets and Specialized Producers

While TAF represented industrial mass production, companies like Wild Hornets exemplified specialized, agile manufacturing focused on innovation and customization. Wild Hornets operated at a smaller scale, hundreds of drones weekly rather than thousands but with a focus on advanced capabilities and rapid prototyping.

The company's facility looked more like a high-tech workshop than a factory. Engineers and technicians worked in small teams, each focused on specific product lines or experimental concepts. The atmosphere emphasized creativity and problem-solving over strict procedures. When combat units requested modifications or entirely new capabilities, Wild Hornets could design, prototype, and produce small batches within days.

Wild Hornets specialized in several niches. The company produced long-range FPV platforms with extended flight times, using custom battery configurations and aerodynamic refinements. They manufactured specialized reconnaissance drones with advanced sensor packages, thermal imaging, zoom optics, laser rangefinders. Some of their platforms incorporated cutting-edge features like AI-assisted navigation or swarm coordination capabilities, experimental systems tested in real combat conditions.

The customer base for Wild Hornets differed from TAF's mass market approach. Elite units, special operations forces, reconnaissance battalions, and specialized drone units came to Wild Hornets for platforms with capabilities beyond standard production models. These customers could articulate specific requirements and provided detailed feedback from combat operations, creating a virtuous cycle of improvement.

Customization was Wild Hornets' signature offering. A unit might request FPV drones optimized for forest operations, requiring specific motor configurations for maneuvering between trees. Another might need platforms with redundant control systems for operation in high-jamming environments. Wild Hornets built to specification, charging premium prices but delivering exactly what combat units needed.

The supply chain strategy differed markedly from TAF's approach. Rather than vertical integration and bulk purchasing, Wild Hornets maintained flexibility through diverse supplier networks and willingness to use premium components. If a specific camera sensor delivered superior performance, Wild Hornets would source it regardless of cost, passing the expense to customers who valued capability over price.

Innovation at Wild Hornets focused on emerging technologies. The company experimented with autonomous navigation systems that could complete missions even if control links were jammed. They developed thermal signature reduction techniques to make drones harder for enemy sensors to detect. Some projects explored entirely new concepts, drones that could perch and observe for extended periods, or platforms designed specifically for urban warfare.

The workforce at specialized manufacturers like Wild Hornets tended toward different backgrounds than mass production facilities. Many employees came from tech startups, bringing software development and rapid prototyping skills. Some were hobbyists from the pre-war drone racing community, intimately familiar with FPV technology. University engineering graduates found positions where they could apply cutting-edge concepts directly to military applications.

Collaboration between specialized producers and larger manufacturers created knowledge diffusion throughout the ecosystem. A successful innovation at Wild Hornets might be adopted by TAF for mass production, scaled from dozens to thousands. Conversely, Wild Hornets sometimes licensed designs from larger manufacturers and created enhanced variants. This horizontal knowledge transfer accelerated innovation across the entire sector.

Specialized manufacturers also served as testing grounds for radical concepts. Because they produced smaller batches, they could afford to fail. Not every experimental platform succeeded, some proved too complex for battlefield conditions, others too expensive even for elite units, still others failed to deliver promised performance. But the willingness to experiment drove progress, and successful innovations eventually diffused throughout the ecosystem.

By 2025, Ukraine had dozens of specialized drone manufacturers, each with particular niches and expertise. Some focused on long-range strike platforms, others on counter-drone systems, still others on reconnaissance. The diversity strengthened the overall ecosystem by ensuring redundancy and encouraging innovation through competition.

The Regulatory Gray Zone: Speed Over Bureaucracy

One of the most remarkable aspects of Ukraine's drone manufacturing ecosystem was what it lacked: bureaucratic oversight, lengthy certification processes, and peacetime regulatory frameworks. The normal rules of defense procurement had been suspended, replaced by an environment that prioritized speed, adaptation, and results over procedural compliance.

In peacetime, a new military drone platform might require years of testing, evaluation, and certification before fielding. Design specifications would be frozen early and changes would trigger lengthy reapproval processes. Safety requirements, environmental regulations, and procurement procedures would all slow the path from concept to combat. Ukraine had none of this.

Instead, the system operated on trust and demonstrated performance. A manufacturer could design a new platform, produce a small batch, and deliver directly to combat units for battlefield testing. If soldiers liked the performance, they would order more often funding purchases through crowdfunding or unit funds rather than official procurement channels. Successful designs scaled up through organic demand rather than centralized decision-making.

The absence of formal certification meant manufacturers bore direct responsibility for quality. There were no third-party testing laboratories or government certification boards. If a manufacturer's drones consistently failed in combat, units stopped ordering them. Reputation became the primary quality control mechanism, with word-of-mouth spreading rapidly through military networks about which manufacturers delivered reliable systems.

This regulatory gray zone enabled remarkable speed. When fiber-optic guided drones appeared on the battlefield in spring 2024, Ukrainian manufacturers reverse-engineered the concept and began production within weeks. There were no patent disputes, no licensing negotiations, no regulatory approvals required. The imperative was simple: if it worked, make more of it.

The system's flexibility extended to modifications and improvements. Combat units often modified purchased drones themselves, swapping motors, changing propellers, or adding equipment. Some units established their own small-scale manufacturing, producing drones for internal use.

This grassroots innovation fed back to manufacturers, who incorporated successful modifications into subsequent production runs.

Intellectual property protection was essentially nonexistent within Ukraine's drone ecosystem. Successful designs were copied freely, with manufacturers openly reverse-engineering competitors' platforms. Rather than slowing innovation, this openness accelerated it, improvements spread rapidly and manufacturers competed on production capacity, quality, and service rather than exclusive design rights.

The relationship between manufacturers and the government evolved throughout the conflict. Initially, the government played minimal role beyond providing funding through various support programs. By 2025, more formal structures were emerging, production contracts, quality standards, and coordination mechanisms but the culture of speed and flexibility remained. Bureaucracy was emerging, but it was bureaucracy designed to support rather than control.

Critics of this approach pointed to legitimate concerns. Without centralized oversight, quality varied significantly between manufacturers. Some production facilities operated with minimal safety standards, leading to occasional accidents. The lack of standardization complicated logistics and training, different manufacturers' drones required different spare parts and sometimes different piloting techniques.

Supporters countered that the approach delivered results that traditional procurement could never match. Ukraine had built a drone manufacturing capacity that exceeded most NATO countries combined, and done so in under three years. The innovation rate was extraordinary, with new capabilities appearing constantly. The cost efficiency was unmatched, Ukrainian drones cost a fraction of Western equivalents while delivering comparable performance.

The regulatory gray zone also enabled rapid scaling of production. Manufacturers didn't need to wait for government approval to expand facilities or increase output. If demand existed and financing was available, expansion proceeded. This resulted in rapid growth of production capacity, though sometimes at the cost of efficiency or organization.

By October 2025, Ukraine was beginning to grapple with transitioning from wartime chaos to sustainable industry. Some regulation was inevitable, safety standards, quality minimums and export controls. The challenge was institutionalizing lessons learned while preserving the speed and flexibility that had enabled success. The balance remained uncertain, with ongoing debates about how much structure was necessary without stifling innovation.

Quality Control in Wartime Conditions

Manufacturing thousands of drones weekly under combat conditions created unique quality control challenges. Traditional manufacturing philosophy emphasized prevention, catching defects before they reached customers. Wartime production emphasized speed, accepting some defects as the price of maintaining production rates. The balance between quality and quantity became a constant negotiation.

Basic quality control began with component inspection. Electronic components, motors, propellers, and frames all underwent visual inspection and basic functionality testing upon receipt. Defective components were set aside, though in periods of shortage, even marginal components might be used with the knowledge that failure rates would increase.

Assembly quality varied by manufacturer. Larger facilities like TAF employed trained assemblers who specialized in specific tasks, producing consistent results. Smaller workshops might have less experienced workers, leading to more variable quality. The learning curve was constant, with new employees making mistakes that experience would eliminate.

Testing protocols represented the crucial quality gate. For FPV kamikaze drones, this typically involved brief test flights in secure areas, verifying that motors spun properly, control responses were crisp, and video transmission was clear. Drones that failed these tests went back for troubleshooting and repair. Successful drones were packaged for shipment.

More sophisticated platforms underwent more extensive testing. Reconnaissance drones might fly complete mission profiles, testing endurance, sensor performance, and navigation systems. Strike quadcopters tested payload release mechanisms and stability under load. Long-range platforms verified engine performance and fuel consumption over extended runs.

The challenge was balancing thoroughness with speed. Comprehensive testing would catch more defects but slow production. Minimal testing maintained production rates but sent defective drones to combat units. Different manufacturers struck different balances based on their production philosophy and customer base.

Combat feedback provided ultimate quality control. Units that received defective drones reported problems, sometimes through formal channels, often through direct communication with manufacturers. Persistent quality issues damaged reputations and reduced orders. This market mechanism enforced minimum quality standards even absent formal oversight.

Some quality issues only appeared in combat conditions. A drone might pass all factory tests but fail when subjected to intensive electronic warfare. Another might perform well in calm weather but struggle in high winds. Combat units learned which manufacturers' products were reliable under specific conditions, information that spread rapidly through military networks.

Batch-to-batch variation presented another challenge. A manufacturer might produce excellent drones for months, then experience quality drops due to component substitutions or workforce changes. Tracking production batches and correlating them with field performance helped identify systematic problems, though this required data collection infrastructure that many manufacturers lacked.

By 2025, quality control was professionalizing. Larger manufacturers employed dedicated quality assurance teams, established documented procedures, and tracked defect rates systematically. Smaller producers remained more informal but increasingly recognized that reputation depended on consistent quality. The market was maturing from its chaotic early days toward something resembling normal industry.

Training Pipelines for Drone Operators

Manufacturing drones was only half the equation, someone had to fly them. The explosive growth in drone numbers required parallel growth in trained operators, creating an enormous training challenge that Ukraine addressed through a combination of formal military schools, volunteer training programs, and distributed on-the-job learning.

Military training centers established dedicated drone operator courses by 2023, teaching basic FPV piloting, quadcopter operation, and reconnaissance drone management. The programs varied in length from two weeks for basic FPV operations to several months for advanced reconnaissance pilots. Training emphasized practical skills, flying in simulators, then transitioning to actual drones in training areas before deploying to combat units.

The curriculum evolved constantly as battlefield lessons accumulated. Early training focused on basic flight skills and drone capabilities. By 2025, training included electronic warfare awareness, tactics for evading air defenses, target identification, and coordination with other military assets. The best training programs incorporated actual combat footage, showing students what to expect and how enemy forces would react.

Simulator training became increasingly sophisticated. Early FPV pilots learned on commercial drone racing simulators, acquiring basic skills before risking actual hardware. By 2025, military-specific simulators replicated combat conditions, jamming, enemy fire, and realistic target environments. Pilots could practice difficult maneuvers repeatedly without consuming drone inventory.

Volunteer organizations filled critical training gaps. Groups like the Angry Birds established training centers that produced pilots faster than formal military programs. These volunteer instructors, often experienced FPV racers or early-war drone veterans, taught in compressed courses emphasizing practical skills over military protocol. The graduates might lack formal military credentials but could fly combat missions immediately.

Distributed training within combat units provided the final polish. New drone operators arrived with basic skills but learned combat tactics from experienced pilots in their units. This apprenticeship model transferred tacit knowledge, understanding jamming patterns in specific sectors, knowing where enemy air defenses were strongest, learning to coordinate with friendly artillery. This knowledge couldn't be taught in schools but was essential for survival and effectiveness.

The background of successful drone operators varied widely. Some came from gaming communities, bringing reflexes and spatial awareness from countless hours in virtual worlds. Others were former pilots of manned aircraft, adapting their aviation knowledge to smaller platforms. Still others were soldiers who volunteered for drone duty, bringing tactical understanding even if they lacked flight experience.

Aptitude testing helped identify promising candidates. Quick reactions, good hand-eye coordination, and spatial reasoning predicted success with FPV platforms. Patience, attention to detail, and systematic thinking suited reconnaissance drone operators. Strike quadcopter pilots needed a mix of both skill sets, precision for bombing runs and patience for loitering operations.

Burnout became a serious concern by 2025. Drone operators flew multiple missions daily, experiencing the stress of combat without the release of physical action. The remote nature of drone operations created psychological challenges, watching targets destroyed through a camera screen, sometimes seeing enemy casualties in graphic detail, then returning to normal life minutes later. Rotation policies and psychological support became necessary to maintain operational effectiveness.

Maintenance and Spare Parts Logistics

For every drone flying combat missions, another dozen sat in maintenance, awaited spare parts, or required repair after battle damage. The logistics of sustaining thousands of drones in operation across hundreds of military units created a supply chain challenge as complex as manufacturing itself.

Spare parts inventory became critical infrastructure. Combat units stockpiled motors, propellers, flight controllers, cameras, and batteries, the components most likely to fail or suffer battle damage. Larger units established mini-workshops where technicians could swap damaged components, test repaired drones, and return them to service within hours.

The supply chain worked on multiple levels. Manufacturers shipped complete drones to distribution centers, along with pallets of spare parts. Regional logistics hubs broke down these shipments and distributed them to combat units based on need and consumption rates. Units requested resupply through digital systems, with urgent requests expedited through priority channels.

Standardization helped enormously where it existed. If multiple manufacturers produced similar FPV designs, spare parts were often interchangeable. A motor from one manufacturer might work in another's frame. This parts compatibility reduced inventory requirements and simplified logistics. Conversely, specialized platforms with unique components created logistics headaches.

Field maintenance ranged from simple to complex. Replacing a damaged propeller required seconds. Swapping a burned-out motor took minutes. Diagnosing and repairing electronic failures might require hours and specialized test equipment. Combat units developed internal expertise, with designated technicians becoming skilled at field repairs.

Some damage was irreparable. Drones that crashed at high speed were often complete losses, with frames shattered and electronics destroyed. Partial battle damage might salvage some components, a drone hit by shrapnel might lose its camera but have serviceable motors. Technicians became expert at scavenging parts from destroyed drones to repair others.

Battery logistics presented unique challenges. Lithium polymer batteries degraded with use and age, requiring constant replacement. Storage and charging procedures required care to prevent fires. Combat units needed multiple batteries for each drone, creating significant inventory requirements. The battery supply chain operated in parallel to drone and spare parts logistics.

Repair facilities ranged from unit-level workshops to centralized depots. Unit-level maintenance handled routine repairs and component swaps. Battalion and brigade maintenance facilities tackled more complex repairs with better diagnostic equipment. Centralized depots refurbished heavily damaged drones, sometimes completely rebuilding airframes around salvaged electronics.

The consumption rate of spare parts tracked with operational tempo. During intensive combat operations, units might exhaust spare parts inventories within days, creating critical shortages that grounded drones until resupply arrived. During quieter periods, parts accumulated in unit inventories. Managing this variation required flexible logistics systems and significant safety stock.

By 2025, the logistics system had matured considerably from early-war improvisation. Digital tracking systems monitored spare parts inventories at all levels. Predictive algorithms forecasted consumption rates and triggered resupply automatically. Mobile maintenance teams could deploy to units experiencing equipment shortages. The system wasn't perfect, shortages still occurred, and some specialized components remained difficult to source but it functioned far better than in 2022-2023.

The 100% Domestic Supply Chain Achievement

Perhaps the most strategically significant achievement of Ukraine's drone manufacturing ecosystem was the nearly complete domestication of the supply chain. By late 2024, Ukrainian manufacturers claimed to produce FPV kamikaze drones using entirely domestic components, no imports required. This supply chain independence represented a remarkable industrial accomplishment with profound strategic implications.

The journey to domestic production required substituting hundreds of imported components with Ukrainian-made alternatives. Electronic components, microcontrollers, voltage regulators and radio transceivers required developing or licensing designs, then establishing production facilities. Carbon fiber frames needed material sourcing and molding capabilities. Motors required precision manufacturing and quality control. Cameras and video transmission systems demanded specialized electronics production.

Some substitutions were straightforward. Carbon fiber composite manufacturing already existed in Ukraine, requiring only adaptation to drone-specific requirements. Electronic assembly capacity could be established relatively quickly using imported manufacturing equipment. Other components proved more challenging, requiring months of development and tooling.

The strategy combined multiple approaches. Where possible, Ukrainian companies licensed existing designs from foreign partners, establishing domestic production under license. For simpler components, reverse engineering yielded functional equivalents quickly. The most critical components received government investment in dedicated production facilities.

The achievement was never quite as complete as claimed. Some specialized electronic components remained difficult to source domestically, particularly cutting-edge semiconductors and precision sensors. But the supply chain was sufficiently robust that external disruption like China's restrictions on DJI drone exports no longer threatened production capacity.

The strategic benefits extended beyond insulation from supply disruptions. Domestic production employed Ukrainian workers, keeping money and expertise within the country. It developed industrial capabilities that would remain valuable after the war. It positioned Ukraine as a potential drone exporter, competing in global markets based on combat-proven designs and low costs.

The cost structure of domestic production proved competitive with imports. Ukrainian labor rates were lower than Western countries, offsetting any economies of scale advantages foreign manufacturers might have. Eliminating import duties and shipping costs reduced per-unit expenses. The high production volumes achieved by 2025 enabled domestic manufacturers to achieve their own economies of scale.

Quality initially lagged behind imported components, but improved rapidly. Ukrainian-made motors initially failed more frequently than imported equivalents, but design improvements and manufacturing refinement closed the gap. By 2025, domestically-produced components often matched or exceeded imported alternatives in reliability.

The ecosystem's resilience was tested repeatedly. Russian strikes targeted manufacturing facilities, destroying some production capacity. Component shortages emerged periodically, creating temporary bottlenecks. Yet the distributed nature of production meant that no single strike could cripple the system, and backup suppliers generally existed for critical components.

By October 2025, Ukraine's drone manufacturing ecosystem stood as testimony to what desperate necessity, entrepreneurial energy, and national mobilization could achieve. From importing consumer drones in 2022, Ukraine had built an industrial base producing over 150,000 drones monthly, employing tens of thousands of workers, and pioneering technologies that were reshaping warfare globally. The ecosystem wasn't perfect, quality varied, bureaucracy was emerging, and challenges remained but it worked. And in wartime, that was enough.

PART V: LESSONS AND IMPLICATIONS

Chapter 21

Tactical Revolution

The proliferation of drones across the Russia-Ukraine battlefield fundamentally transformed how small units fought, moved, and survived. By 2025, tactics that had worked for decades, even centuries became obsolete, replaced by new doctrines written in real-time by soldiers adapting to the reality of constant aerial observation. The tactical revolution touched every aspect of ground combat, from how infantrymen crossed open ground to how artillery chose firing positions, from the utility of camouflage to the very concept of the battlefield's geography.

How Drones Changed Combined Arms Warfare

Combined arms warfare, the coordinated employment of infantry, armor, artillery, and aviation had dominated military thinking since World War I. The fundamental principle was simple: different arms supported each other, with each compensating for others' weaknesses. Infantry protected tanks from enemy infantry. Tanks provided infantry with mobile firepower and protection. Artillery suppressed enemy positions before assaults. Aviation provided reconnaissance and strike capability.

Drones didn't replace these traditional arms, but they fundamentally altered the relationships between them. The changes manifested across every phase of operations, from planning through execution to exploitation.

Reconnaissance, traditionally provided by dedicated scout units or aerial platforms, became ubiquitous and immediate. Every infantry company operated its own reconnaissance drones, providing commanders with real-time video of enemy positions. The fog of war didn't disappear, electronic warfare and weather still limited visibility but it thinned dramatically. Commanders planning attacks could see enemy trenches, count fighting positions, and identify weak points without exposing scouts to enemy fire.

This reconnaissance revolution compressed planning timelines. In previous conflicts, gathering intelligence for an attack might require days of patrol activity, prisoner interrogation, and analysis. By 2025, a commander could launch a reconnaissance drone, survey objective areas within hours, and plan attacks based on current information rather than dated intelligence. The operational tempo accelerated accordingly.

Artillery employment transformed most dramatically. Traditional artillery relied on forward observers, soldiers positioned where they could see targets to direct fire. Observers radioed coordinates to gun positions kilometers away, then watched shells impact and called corrections. The process was cumbersome, requiring clear lines of sight and exposing observers to enemy fire.

Drone-directed artillery replaced forward observers with aerial spotters that could see deeper, survive longer, and adjust fire with precision that human observers rarely matched. An artillery unit could now receive target coordinates from reconnaissance drones, fire missions, and have strike quadcopters or FPV drones assess damage immediately. Some artillery units integrated drones so thoroughly that gun crews never saw their targets with human eyes, working entirely from drone video feeds.

The accuracy improvements were substantial. Artillery guided by drone spotters achieved first-round hit rates that previously required multiple ranging shots. This saved ammunition which was increasingly scarce for both sides and reduced the time targets had to seek cover or relocate. The cumulative effect was artillery that hit harder, faster, and more efficiently than ever before.

Armor operations faced the most dramatic tactical changes. Tanks and armored vehicles, dominant on battlefields since World War I, suddenly became vulnerable to weapons costing a fraction of their value. An FPV drone carrying a shaped charge could destroy a multimillion-dollar tank, and reconnaissance drones could spot armored vehicles from safe distances, vectoring in FPV strikes or artillery.

Armored units adapted through dispersion, speed, and combined arms integration. Tank platoons that once moved in close formation now dispersed over hundreds of meters, reducing the chances that a single reconnaissance drone could spot the entire unit. Movement happened rapidly, with vehicles spending minimal time exposed. Most critically, armored units never operated without their own drone protection, reconnaissance drones scouting ahead, FPV teams ready to strike enemy drone operators, and electronic warfare systems attempting to jam enemy drones.

The integration of drones into combined arms teams created new organizational structures. By 2025, Ukrainian mechanized battalions typically included dedicated drone platoons with reconnaissance, strike, and FPV capabilities. These drone elements worked alongside traditional arms, providing aerial fires and intelligence that complemented ground maneuver and artillery support.

Offensive operations incorporated drones into every phase. Before an attack, reconnaissance drones mapped enemy positions and identified priority targets. As the assault began, FPV drones struck defensive positions, suppressing enemy fires while friendly infantry advanced. Strike quadcopters attacked enemy reserves and artillery positions, preventing reinforcement. During exploitation, reconnaissance drones tracked retreating enemy forces and identified counterattack preparations.

Defensive operations similarly integrated drones throughout. Reconnaissance platforms provided early warning of enemy buildups. Strike drones engaged advancing forces, particularly armored vehicles. FPV teams hunted enemy observation posts and artillery spotters. The layered defense now extended vertically, with aerial elements as integral as trenches and minefields.

Infantry Tactics in the Age of Constant Aerial Observation

For the individual infantryman, drones transformed the fundamental calculus of survival. The assumption that had governed infantry tactics for generations, that small units could move undetected between covered positions no longer held. Any movement might be observed by enemy drones, and observation could quickly translate to artillery fire or FPV strikes.

Movement techniques evolved rapidly. The traditional infantry advance, squads moving by bounds from cover to cover became suicidal in areas under drone observation. Instead, movement happened under specific conditions designed to reduce drone detection probability: during poor weather that grounded drones, at night when electro-optical cameras struggled, immediately after friendly jamming pulses disabled enemy control systems, or under smoke screens that obscured aerial observation.

When movement was unavoidable during daylight and good weather, infantry adopted tactics borrowed from guerrilla warfare. Small groups moved separately, never concentrating in vulnerable masses. Movement routes exploited tree cover, ravines, and built-up areas where overhead observation was limited. Soldiers moved quickly between covered positions, minimizing time in the open. Units accepted dispersion as necessary, even though it reduced immediate combat power.

The concept of the "safe rear area" effectively disappeared. Drones could strike anywhere their range permitted, which for long-range platforms meant dozens of kilometers behind front lines. Rest areas, supply points, and command posts all required dispersal and camouflage. Soldiers learned that concentration invited attack, whether from artillery guided by reconnaissance drones or direct strikes by FPV platforms.

Patrolling, a basic infantry task, became extraordinarily dangerous. Traditional patrols moved in small groups, typically four to ten soldiers seeking to observe enemy positions while avoiding detection. Drones inverted this equation. Enemy reconnaissance drones could observe patrols from safe distances, and FPV drones could strike them with minimal warning. Ukrainian and Russian patrols suffered horrific casualties in 2023 before tactics adapted.

The adaptations included smaller patrol sizes, sometimes just two soldiers and greater reliance on stealth over firepower. Patrols moved at night almost exclusively, used electronic warfare detectors to identify enemy drone activity, and often carried their own small quadcopters to search ahead. Some patrols abandoned traditional patrol routes entirely, instead positioning observers in hidden locations to watch specific areas without movement.

Close combat changed less dramatically, as drones struggled to distinguish friend from foe when infantry closed to contact. But even here, drones influenced tactics. Attacking infantry launched FPV strikes on defensive positions immediately before assaulting, suppressing defenders during the critical final approach. Defenders used strike quadcopters to drop munitions on attacking forces, particularly when attackers clustered in assembly areas before final assaults.

The psychological burden on infantry was immense. The knowledge that enemy eyes might be watching at any moment, that death could arrive from the sky without warning, created constant stress. Soldiers developed near-paranoid attention to aerial sounds, diving for cover at the buzz of propellers. The mental strain contributed to combat fatigue even among units not actively engaged in fighting.

Training had to prepare infantry for this new reality. Recruits learned to identify drone types by sound and appearance, practiced immediate action drills for drone attacks, and trained in camouflage techniques specifically designed to defeat aerial observation. The curriculum included electronic warfare basics—understanding how jamming worked and how to operate when drones disrupted GPS and communications.

The Death of the "Safe Rear Area"

The traditional battlefield had clear geography: a front line where combat occurred, rear areas where support functions operated in relative safety, and deep rear zones almost entirely secure. This spatial organization allowed military forces to position vulnerable assets, field hospitals, supply depots, maintenance facilities and command posts at distances where enemy fires couldn't reach them.

Drones, particularly long-range strike platforms, obliterated this comfortable geography. By 2025, Russian Shahed drones struck targets across all of Ukraine, including cities like Lviv near the Polish border, more than 1,000 kilometers from the nearest Russian-controlled territory. Ukrainian long-range drones hit targets deep in Russia itself, attacking oil refineries in Tatarstan and military facilities in regions that had never experienced direct attack.

The implications cascaded through military operations. Support units that had operated with relative impunity now required the same security measures as front-line combat forces. Field hospitals dispersed patients across multiple small facilities rather than concentrating in large medical complexes. Supply depots scattered inventory across numerous small caches instead of maintaining centralized warehouses. Command posts moved frequently and operated from camouflaged or hardened positions.

The logistics burden increased substantially. Dispersal meant more facilities to guard, more transportation required to move supplies between scattered locations, and more communications infrastructure to coordinate distributed operations. Units devoted greater resources to force protection, air defense systems, camouflage, and defensive positions, resources that couldn't be used for offensive operations.

Civilian-military distinction blurred as drones struck rear-area cities. Russian drone attacks on Ukrainian cities like Odesa, Kharkiv, and Kyiv killed civilians while targeting military and industrial facilities. The constant threat transformed civilian life, with millions of Ukrainians spending nights in shelters and entire cities learning to function under recurring air raids. The economic impact was enormous, with businesses disrupted, infrastructure damaged, and investors reluctant to commit resources to locations under regular attack.

Personnel rotation became more dangerous. Moving soldiers from front-line units to rear areas for rest had always been part of military operations, allowing troops to recover before returning to combat. But when rear areas faced drone strikes, the rest was interrupted by air raid sirens and the real risk of casualties. Some soldiers reported that "rest" periods in rear cities felt more stressful than front-line service, as they worried about families and felt helpless against drone attacks.

Training facilities had to adapt to the elimination of safe rear areas. Pre-war training typically occurred far from combat zones, where soldiers could learn without enemy interference. By 2025, Ukrainian training centers operated under the same threat as combat units, with air defense systems protecting ranges and trainees learning to respond to air raid warnings as part of basic training.

The strategic implications were profound. Traditional military planning assumed that forces could withdraw from front lines to reconstitute, absorb replacements, and prepare for future operations. When no safe reconstitution areas existed, this assumption failed. Units remained under threat even when not engaged in active combat, accumulating stress and preventing full recovery.

Commanders had to weigh risks constantly. Concentrating forces for offensive operations created lucrative targets for enemy drones. But excessive dispersion reduced combat power and complicated command and control. The balance shifted based on enemy drone activity, weather conditions, and operational necessity, requiring constant adjustment.

Trenches and Camouflage in the Drone Age

The return of trench warfare to European battlefields surprised many observers who expected modern warfare to emphasize mobility and maneuver. But the combination of artillery, drones, and precision strike created conditions remarkably similar to World War I where exposure meant death and survival required digging in. The difference was that World War I trenches were designed to protect against artillery and direct fire. By 2025, trenches also needed to defeat aerial observation.

Trench design evolved rapidly. Early trenches followed traditional designs, linear positions connected by communication trenches, with overhead cover in some positions. These proved vulnerable to drone-directed artillery, which could identify trenches from above and call precise strikes. Soldiers died in positions that would have protected them from ground-level threats.

The evolution emphasized camouflage and deception. Trenches were routed under tree cover wherever possible, making aerial observation more difficult. Communication trenches followed irregular paths rather than straight lines, breaking up signatures visible from above. Overhead camouflage netting covered positions, making them harder to distinguish in drone video feeds.

Some units built elaborate false positions, dummy trenches and fighting positions designed to attract enemy attention and waste ammunition. These deception efforts required significant labor but sometimes succeeded in drawing FPV strikes and artillery fire away from actual defensive positions. The effectiveness varied; experienced drone operators learned to distinguish occupied positions from decoys through subtle indicators like worn paths, fresh earth, or heat signatures.

Entrances and exits to fighting positions received particular attention. Drone operators learned to watch for soldiers entering and exiting bunkers, identifying occupied positions. Defenders responded by creating multiple entrances, using them irregularly to confuse observation, and sometimes building false entrances to occupied positions while using concealed access points.

The depth and overhead protection of positions increased throughout the conflict. Early trenches often had minimal overhead cover, adequate against small arms and shell fragments but vulnerable to direct strikes. By 2025, front-line positions featured multiple layers of overhead protection, logs, earth, sandbags, sometimes concrete designed to survive direct hits from drone-dropped munitions and FPV strikes.

Underground positions became increasingly common, particularly for command posts and ammunition storage. These positions, sometimes dug several meters deep, offered protection from everything short of the largest munitions. The excavation required enormous labor and engineering resources, but the protection justified the investment. Some positions incorporated multiple rooms, ventilation systems, and hardened access points, resembling miniature fortresses.

Camouflage evolved into sophisticated science. Military units employed camouflage specialists who studied how positions appeared in drone video feeds and designed countermeasures. This included matching natural vegetation, breaking up straight lines and geometric patterns, and using thermal insulation to reduce heat signatures visible to infrared cameras.

The temporal dimension of camouflage became crucial. Fresh earth from excavating positions showed clearly in aerial imagery, identifying recent construction. Defenders learned to camouflage positions immediately during construction, spreading removed earth over wide areas and transplanting vegetation to cover disturbed ground. Some positions were built months before occupation, allowing vegetation to recover and positions to blend into terrain.

Movement between positions required careful planning. Worn paths visible from above marked routes between fighting positions and rear areas, allowing enemy drones to predict where soldiers would appear. Units developed multiple paths, varied their routes, and sometimes suspended camouflage netting over heavily-used routes to prevent aerial observation.

The interaction between natural and artificial camouflage created constant tactical choices. Natural cover, forests, ravines and built-up areas provided excellent concealment but constrained positioning and limited fields of fire. Artificial camouflage allowed greater positioning flexibility but required maintenance and could fail if not properly maintained. Most units combined both approaches, using natural cover where available and supplementing with artificial camouflage.

Seasonal variation affected camouflage effectiveness dramatically. Summer vegetation provided excellent concealment, with full tree canopies and ground cover obscuring positions. Winter snow revealed even subtle traces of activity, paths, disturbed snow around positions, heat signatures more visible against cold backgrounds. Spring and autumn represented intermediate conditions, with partial vegetation and variable weather.

Electronic Warfare as Essential Capability

Electronic warfare, the use of electromagnetic spectrum to attack enemy systems while protecting friendly ones evolved from specialized capability to basic necessity. By 2025, no unit operated without electronic warfare considerations, and dedicated electronic warfare systems proliferated down to company and even platoon level.

The primary electronic warfare mission was countering enemy drones through jamming. Jamming systems broadcast radio interference on frequencies used for drone control and video transmission, disrupting enemy operators' ability to fly their platforms. Early jamming systems were bulky and power-hungry, mounted on vehicles or deployed in fixed positions. By 2025, portable jammers small enough for individual soldiers to carry had become common.

The effectiveness of jamming varied enormously based on frequency, power, and drone sophistication. Simple commercial drones using standard frequencies were relatively easy to jam, their control signals overwhelmed by jamming interference. Military drones with frequency-hopping or spread-spectrum capabilities resisted jamming much better, requiring more powerful or sophisticated jammers to disrupt.

The tactical employment of jamming required careful consideration. Active jammers broadcast powerful radio signals, creating signatures that enemy electronic warfare systems could detect and locate. Activating jamming might protect against enemy drones but simultaneously reveal positions for enemy artillery strikes. Commanders constantly weighed these risks, activating jamming during critical moments but maintaining radio silence at other times.

GPS jamming and spoofing became important elements of the electronic warfare fight. Both sides deployed systems that either jammed GPS signals preventing drones from using satellite navigation or spoofed them, broadcasting false GPS signals that caused drones to believe they were in different locations than actual. These techniques were particularly effective against long-range platforms that relied heavily on GPS navigation.

The response to GPS jamming drove innovation in alternative navigation. Drones incorporated inertial navigation systems including gyroscopes and accelerometers that tracked movement allowing continued operation after GPS loss. Some platforms used terrain-matching algorithms, comparing visual imagery to stored maps to determine location. The fiber-optic controlled drones that emerged in 2024 bypassed GPS entirely, guided through direct cable connections immune to jamming.

Counter-countermeasures proliferated. As jamming became ubiquitous, drone designers incorporated anti-jamming features. Frequency-agile radios automatically switched to less congested frequencies when jammed. Directional antennas focused signals toward control stations, requiring less power and reducing susceptibility to broad-spectrum jamming. Some advanced drones incorporated autonomous flight capabilities, continuing missions even when control links were disrupted.

Detection systems complemented jamming capabilities. Radio frequency sensors detected enemy drone control signals, alerting defenders to drone presence and sometimes providing direction-finding data that located enemy control stations. Acoustic sensors listened for propeller noise, detecting approaching drones before they came into view. Radar systems, ranging from vehicle-mounted military radars to adapted commercial marine radars, detected drones at distance.

The layered integration of detection and jamming created localized air defense bubbles. A typical Ukrainian defensive position by 2025 might include radar detection, radio frequency sensors for technical intelligence, acoustic sensors for close-range warning, and multiple jammers covering different frequency bands. Operators monitored these systems continuously, activating jammers when threats appeared and attempting to locate enemy control stations for counter-strike.

Electronic warfare extended beyond drone operations. Both sides jammed enemy communications, disrupting coordination and command and control. Counter-battery radars systems that detected artillery fire and calculated firing positions became priority targets for jamming and destruction. Navigation systems, tactical data links, and even civilian cell phone networks became battlegrounds in the electromagnetic spectrum.

The personnel requirements for electronic warfare increased dramatically. Pre-war, electronic warfare was a specialized field requiring extensive training. By 2025, every combat unit needed soldiers who understood basic electronic warfare principles, could operate jamming equipment, and recognized electronic attack indicators. Training pipelines struggled to produce sufficient trained personnel, and much expertise developed through on-the-job learning.

The dynamic nature of electronic warfare created constant adaptation cycles. A jamming technique would prove effective, enemy drones would adapt, new jamming methods would emerge in response. The cycle time, measured in weeks or months, created a perpetual arms race in the electromagnetic spectrum that paralleled the physical battlefield.

Small Unit Tactics Transformed

At the level where soldiers actually fought, squads, platoons, and companies the tactical changes were most immediate and personal. Leaders who had trained for decades in traditional tactics found their knowledge obsolete, replaced by doctrines developed through trial, error, and brutal casualties.

Squad movement, the foundation of infantry tactics, became exercises in minimizing exposure. The traditional squad formation, teams moving by bounds while others provided covering fire, presumed that observation was intermittent and localized. Drones made observation constant and wide-ranging. Squads adapted by moving only when necessary, exploiting brief windows when drone surveillance was disrupted by jamming or weather, and accepting greater dispersion even though it reduced immediate mutual support.

Fire and maneuver, the basic tactical method for attacking enemy positions, required fundamental modification. The traditional approach had one element fixing the enemy with suppressive fire while another maneuvered to attack from a different direction. But the maneuvering element, necessarily exposed during movement, became vulnerable to drone-directed fires. Some units developed techniques where FPV drones provided the fixing element, striking or threatening enemy positions while infantry maneuvered. Others used smoke extensively, creating visual obscuration that defeated drone observation.

Hasty attacks, rapid assaults on encountered enemy positions became extraordinarily difficult. Traditional hasty attack doctrine emphasized speed and violence of action, with infantry immediately assaulting before enemies could organize effective defense. But speed meant exposure, and exposure to drone observation often resulted in devastating fires before infantry could close with objectives. Units adapted by conducting more deliberate attacks even when time was limited, using reconnaissance drones to gather intelligence and FPV strikes to suppress positions before committing infantry.

Defensive tactics proved more adaptable to the drone environment. Static defensive positions could be camouflaged and hardened against aerial observation. Defenders enjoyed advantages in that they could prepare positions carefully, incorporate overhead cover, and had detailed knowledge of local terrain. Nevertheless, even defensive positions required constant adaptation as enemy reconnaissance drones probed for weaknesses and directed fires against identified positions.

The concept of key terrain, geographic features that provided tactical advantage expanded to include electromagnetic terrain. Areas with natural jamming, cities with dense wireless infrastructure and valleys where radio propagation was limited became valuable because they complicated enemy drone operations. Conversely, areas with clear line-of-sight to control stations and minimal electronic interference became dangerous for their susceptibility to drone strikes.

Night operations increased in importance as units sought to exploit the reduced effectiveness of electro-optical drone cameras. Thermal imaging partially compensated, but many drones lacked thermal capabilities, and even those that had them found detection more difficult in darkness. Units conducted more operations at night than they had in peacetime training, requiring adaptation in navigation, coordination, and fire control.

Small unit leadership required new skills and knowledge. Squad and platoon leaders needed to understand drone capabilities and limitations, recognize when their units were under observation, and employ counter-drone tactics. They coordinated with attached drone teams, integrating aerial reconnaissance and strikes into small unit operations. The cognitive load on junior leaders increased substantially, with more factors to consider and more assets to coordinate than traditional infantry operations.

The integration of drone operators into small units created organizational challenges. Where did drone teams fit in squad and platoon structures? How were they commanded, organic to infantry units or attached from specialized drone platoons? Who controlled their employment, infantry commanders or drone specialists? Different units adopted different solutions, but successful approaches generally emphasized integration, with drone operators embedded in infantry units and operating under infantry command while maintaining technical expertise.

Training for the new tactical reality required extensive modification of traditional programs. Recruits learned drone awareness alongside basic infantry skills. Small unit leader courses incorporated drone employment and counter-drone tactics. Collective training exercises included drone threats, requiring units to defend against aerial observation and strike. The training overhead increased substantially, with more skills required and less time available to develop them.

By October 2025, the tactical revolution was ongoing rather than complete. New techniques continued to emerge, driven by battlefield experience and technological evolution. What was clear was that the tactics of 2022 let alone those of earlier conflicts were obsolete. The drone age had rewritten the rules of small unit combat, creating a tactical environment more complex, more lethal, and more demanding than any soldiers had previously faced. Those who adapted survived. Those who didn't became casualties, their deaths hard lessons in the unforgiving school of combat innovation.

Chapter 22

The Economics of Drone Warfare

The mathematics of modern warfare have been fundamentally rewritten on the battlefields of Ukraine. In the spring of 2024, a Ukrainian air defense battery faced a dilemma that would have seemed absurd to military planners just three years earlier. A swarm of Russian drones approached their position, each costing approximately $500 to manufacture. The battery commander watched his screens, knowing that each interceptor missile in his arsenal carried a price tag of $5 million. The equation was brutally simple: even a perfect kill ratio would represent an economic catastrophe.

This moment encapsulates the central paradox of drone warfare that has emerged from the Russia-Ukraine conflict. The side attacking with cheap, expendable systems forces its opponent into an economically unsustainable defensive posture. It is asymmetric warfare taken to its logical extreme, where the very act of successful defense accelerates economic defeat.

The Cost Revolution

To understand the economic transformation, we must first grasp the radical cost compression that drones represent. A typical Ukrainian FPV kamikaze drone rolls off production lines for between $400 and $800, depending on specifications and payload. Russian equivalents cost similar amounts. These are not precision instruments built in sterile facilities by aerospace engineers. They are assembled in workshops, garages, and small factories by technicians who learned their trade during the war itself.

Compare this to the weapons systems they have effectively replaced or supplemented. A single precision-guided artillery shell costs $5,000 to $10,000. A Javelin anti-tank missile runs $175,000 per unit. An air-launched cruise missile can cost $1 million or more. Even unguided artillery shells, the cheapest form of traditional firepower, cost $400-$800 each when accounting for propellant charges and logistical overhead.

The FPV drone matches or exceeds these systems in several critical metrics. It provides real-time visual confirmation of the target. The operator can abort the mission up until the final second before impact. It can maneuver around obstacles, pursue moving targets, and even strike concealed positions that indirect fire cannot reach. And it does all this at a fraction of the cost.

Volodymyr Dubovik, a logistics officer with Ukraine's 93rd Mechanized Brigade, explained the calculation in stark terms during an interview in September 2024: "We can produce ten drones for the cost of one artillery shell. Those ten drones can achieve what would require fifty shells in the old way of fighting. The economics are not just favorable, they are revolutionary."

The production costs tell only part of the story. Traditional weapons systems carry enormous logistical tails. Artillery requires trucks to move shells, specialized vehicles to transport guns, maintenance crews, and secure ammunition depots. Aircraft need runways, fuel supply chains, trained maintenance personnel, and sophisticated support infrastructure. Missiles demand climate-controlled storage, specialized handling equipment, and complex launch systems.

Drones, by contrast, arrive at the front in cardboard boxes. A single pickup truck can carry fifty FPV drones, enough firepower to destroy dozens of targets. They require no special storage conditions. Operators can be trained in weeks rather than months or years. When a drone is lost, it represents less than one day's salary for a typical soldier in a Western military. The entire logistical paradigm shifts from feeding a hungry industrial beast to distributing simple, lethal packages.

The Defender's Dilemma

The economic advantage flips dramatically when examining the defender's perspective. Ukraine has developed one of the most effective integrated air defense systems in modern warfare, regularly achieving interception rates of 90-93% against incoming drone swarms. This is a remarkable technical achievement. It is also economically catastrophic.

The systems shooting down $500 drones employ a mix of technologies, none of them cheap. Patriot missiles, used against larger drones and cruise missiles, cost $4 million each. Even shorter-range systems like the German Gepard anti-aircraft gun fire 35mm rounds costing $30-$50 per shot. When a Gepard expends 200 rounds shooting down a single Shahed drone, a typical engagement the ammunition cost alone approaches $10,000.

More sophisticated systems present even starker mathematics. A single S-300 surface-to-air missile, part of Ukraine's Soviet-inherited arsenal, costs approximately $1.5 million. NASAMS interceptors run $1-2 million per missile. The newest Western systems being deployed carry similar or higher price tags. Against a $20,000 Shahed-136, this represents a cost exchange ratio of 75:1 or higher. Against smaller FPV drones, the ratio becomes absurd.

The economic burden extends beyond the missiles themselves. Each air defense system requires trucks, radars, command posts, trained crews, and maintenance infrastructure. A single Patriot battery demands over 90 personnel and costs approximately $1 billion to acquire. The annual maintenance and operation costs run to tens of millions of dollars. All of this expensive capability must be mobilized to defeat attacks that cost the enemy a tiny fraction of the defensive expenditure.

Maxim Sheremet, an economic analyst with Ukraine's Ministry of Defense who spoke to foreign journalists in August 2024, framed the challenge bluntly: "They are forcing us to play a game where we must spend one thousand dollars to stop one dollar of their attack. The mathematics of this are obvious. No nation can sustain this exchange indefinitely, no matter how wealthy or how committed its allies are."

The summer of 2024 illustrated this dynamic with painful clarity. In July alone, Russia launched over 6,000 drones into Ukraine. Even assuming Ukraine intercepted 90% of these, an optimistic estimate the cost of defensive munitions expended likely exceeded $500 million for a single month. Meanwhile, Russia's expenditure on the drones themselves was probably less than $5 million. The economic exchange ratio approached 100:1 in the attacker's favor.

This unsustainable dynamic forces defenders toward alternative solutions, each with its own economic implications. Electronic warfare systems can jam or spoof drone guidance at much lower cost-per-engagement, but they require expensive initial investment and sophisticated operators. Directed energy weapons such as lasers and high-powered microwaves promise costs measured in pennies per shot, but the systems themselves cost hundreds of millions to develop and deploy. Counter-drone drones offer better cost ratios but introduce new complexity and logistics requirements.

Mass Production as Military Strategy

The economic logic of drone warfare leads inexorably toward mass production. When individual systems are cheap and expendable, victory goes to whoever can produce them in greater numbers. This represents a return to industrial warfare paradigms not seen since World War II, when aircraft, tanks, and ships were produced in thousands rather than dozens.

Ukraine's trajectory illustrates this evolution. In early 2022, the country had almost no domestic drone production. By the end of 2023, it was manufacturing several thousand drones monthly. Throughout 2024, production scaled exponentially. By October, Ukraine produced 1.6 million drones over the course of the year. President Zelensky announced plans to reach 4 million annual production capacity. By early 2025, actual production capacity exceeded 2.5 million units per year, with TAF Drones alone producing 40,000 drones monthly, an operation valued at over $1 billion annually.

These numbers reflect a conscious strategic choice. Ukrainian military planners recognized that they could not win an artillery duel with Russia, which possessed larger stocks and greater production capacity for traditional munitions. But drones offered a different path. The capital investment required to establish drone production was orders of magnitude lower than building artillery factories or aircraft plants. The production cycle was measured in days rather than months. Quality control, while important, could be managed in ways impossible for complex precision weapons.

The distributed production model that emerged solved multiple problems simultaneously. Small manufacturers, some operating from garages and apartments, could produce dozens or hundreds of drones weekly. This distributed approach made production resilient to Russian strikes. It also avoided bottlenecks inherent in centralized manufacturing. Combat units began producing their own drones, customized to their specific tactical needs. The absence of central planning, normally a weakness, became a strength in this context. Innovation happened rapidly, with successful designs spreading through informal networks while failures were quickly abandoned.

The economic implications extended beyond simple unit costs. Traditional defense procurement is notoriously expensive, with byzantine regulations, extensive testing protocols, and multi-year development cycles. A new infantry fighting vehicle might take a decade from concept to deployment, with costs spiraling into billions. Drone development operated on completely different timelines and budgets. A new design could go from concept to front-line deployment in weeks. Total development costs might be $50,000. If the design failed, the financial loss was trivial.

This procurement agility created what economists call "rapid iteration cycles." Manufacturers could afford to experiment because failures were cheap. Combat feedback flowed directly to producers, often the same day a new design was tested in battle. Successful innovations spread virally through the manufacturing network. Within months of Russia introducing fiber-optic guided drones in spring 2024, Ukrainian manufacturers had reverse-engineered the concept and were producing their own versions at scale.

Russia followed a parallel path, though with different organizational structures. The Russian defense industry, more centralized than Ukraine's guerrilla manufacturing network, nonetheless achieved comparable production numbers by late 2023. Initially dependent on Iranian imports of Shahed-136 drones, Russia acquired manufacturing technology and established domestic production. By 2024, Russian facilities produced thousands of Shaheds monthly, supplemented by massive FPV drone production using Chinese components and smuggled Western electronics.

The industrial mobilization on both sides created an arms race defined not by technological sophistication but by manufacturing capacity. The side that could produce more drones could sustain higher loss rates, maintain more persistent surveillance, and overwhelm enemy defenses through sheer numbers. This was warfare as industrial attrition, where factories mattered more than fighter pilots.

The Arithmetic of Attrition

The economic logic of drone warfare reshapes strategic calculations about attrition. In traditional conflicts, military planners carefully husband expensive assets. Losing a $150 million fighter jet is a strategic setback.

Losing a $10 million tank is a significant tactical blow. These losses accumulate slowly because the assets are valuable and relatively scarce.

Drone warfare operates by different rules. An FPV drone has an expected operational lifetime measured in missions, often just one. This is not a flaw but a feature. The entire concept revolves around expendability. You send drones on missions where you expect most or all of them to be lost. The question is not whether you can afford the losses, but whether you can produce replacements faster than you expend them.

This creates a new kind of attrition calculus. Imagine a battalion operating 1,000 drones with a monthly attrition rate of 800 units. If the battalion can acquire 1,000 replacement drones monthly, it maintains capability. If replacement drops to 600, capability degrades rapidly. The mathematics are brutal and simple. Industrial capacity directly translates to combat power in ways that have not been true since World War II.

The July 2024 statistics illustrate this dynamic. Russia launched over 6,000 drones in a single month. Ukraine shot down approximately 90% of them, a defensive success by any reasonable measure. But Russia simply continued producing and launching. The next month brought 5,600 more drones. The month after that, even more. The Russian strategy was not to achieve a favorable exchange ratio in any single engagement, but to force Ukraine into an unsustainable defensive expenditure over time.

Ukraine faced a stark choice: either develop cheaper defensive solutions, or achieve such overwhelming production capacity that it could shift from defense to offense. The Ukrainian response involved both. On defense, Ukraine accelerated development of electronic warfare systems, counter-drone drones, and other lower-cost solutions. On offense, Ukraine ramped up production to launch its own deep-strike campaigns against Russian infrastructure.

By August 2025, Ukraine was conducting sustained long-range drone offensives targeting Russian oil refineries, military-industrial facilities, and logistics hubs hundreds of kilometers inside Russia. These attacks, using more expensive long-range platforms, followed the same economic logic. A $5,000 drone that destroyed $50 million in refinery infrastructure represented a 10,000:1 return on investment. The economics justified significant defensive expenditures by Russia to stop such attacks.

The attrition arithmetic extended beyond hardware to human costs. Training a fighter pilot costs millions of dollars and takes years. Training a drone operator takes weeks and costs thousands. When experienced operators are lost to combat stress, burnout, or enemy action they can be replaced relatively quickly. This commodification of warfare skills has disturbing implications, but economically it represents another form of cost compression.

Implications for Future Conflicts

The economic lessons from Ukraine ripple outward, reshaping how military planners worldwide think about future conflicts. The fundamental insight is simple but profound: in any future conflict involving sophisticated opponents, cheap attack will overwhelm expensive defense unless the defensive technologies undergo equally radical cost reduction.

Consider the implications for naval warfare. A modern destroyer costs $2-3 billion and carries a crew of 300. Its defensive systems are optimized for threats like anti-ship missiles, each costing $1-2 million. Now imagine that same destroyer facing swarms of autonomous boats or aerial drones, each costing $5,000 and carrying enough explosives to mission-kill the ship. The destroyer's defensive systems would be overwhelmed long before the ship ran out of targets. The economics favor the attacker so overwhelmingly that entire classes of expensive platforms become vulnerable.

Air forces face similar recalculations. A modern fighter jet costs $80-100 million. It is defended against by surface-to-air missiles costing $1-3 million. But what happens when that same aircraft faces coordinated swarms of AI-enabled kamikaze drones costing $50,000 each? The fighter cannot carry enough munitions to defeat the swarm. The economics break the traditional air combat paradigm.

Ground forces operating in drone-saturated environments face different but equally challenging mathematics. Armored vehicles costing millions of dollars each become vulnerable to $500 FPV drones. Infantry formations face constant surveillance from quadcopters costing $1,000. The economic cost of protecting these forces escalates dramatically, while the cost of attacking them plummets.

The Ukraine experience suggests several potential responses, each with significant economic implications:

Radical cost reduction in defensive systems. Directed energy weapons promise to bring defensive costs down to match offensive economics. A laser system might fire shots costing pennies in electricity, making it economically feasible to engage thousands of targets. However, the systems themselves remain expensive to develop and deploy. Whether they can be produced in sufficient numbers to matter remains an open question.

Autonomous defensive systems. Counter-drone drones and AI-enabled defensive systems could potentially match the cost profile of attacking drones. If a $1,000 defensive drone can intercept a $1,000 attacking drone, the economics balance. This drives both sides toward autonomous systems and artificial intelligence, whoever achieves superior autonomous capabilities gains decisive advantage.

Restructuring force composition. Expensive platforms designed for previous eras may become obsolete, replaced by larger numbers of cheaper, more expendable systems. Instead of a few exquisite fighters, air forces might field hundreds of loyal wingman drones. Instead of heavy tanks, armies might deploy swarms of small autonomous ground vehicles. This represents a fundamental restructuring of military forces around economic efficiency rather than technological superiority.

Industrial capacity as strategic capability. The Ukraine war demonstrates that the ability to rapidly scale production of military systems is itself a critical strategic capability. Nations that maintain robust manufacturing capacity, flexible supply chains, and rapid innovation cycles possess advantage regardless of their peacetime arsenal size. This shifts strategic planning from stockpiling weapons to maintaining production capacity.

The Lesson We Cannot Unlearn

Perhaps the most profound economic insight from the drone war is this: we have crossed a threshold that cannot be uncrossed. The knowledge of how to build effective military drones from commercially available components is now widespread. The production techniques are documented and disseminated. The costs are low enough that non-state actors, terrorist organizations, and small nations can acquire significant capabilities.

This democratization of precision strike capability fundamentally alters the economics of warfare. Historically, sophisticated military capabilities required nation-state resources to develop and field. Only countries with substantial defense budgets could project power effectively. Drones shatter this paradigm. A crowdfunded volunteer organization in Ukraine can produce thousands of militarily effective systems. If such groups can do it, so can any moderately competent adversary anywhere.

The implications extend beyond conventional warfare. A terrorist organization that once might have struggled to build a single truck bomb can now potentially acquire or manufacture dozens of kamikaze drones. A regional insurgency that previously lacked air power can now conduct aerial attacks. The economic barriers to entry for aerial warfare have collapsed.

Military establishments worldwide are grappling with these implications. Western nations that spent decades optimizing for exquisite, expensive systems are now watching as Ukraine demonstrates the effectiveness of cheap and numerous. The procurement systems that made sense when developing $100 million fighters look increasingly obsolete when the threat is $500 drones produced in garages.

The economic revolution in warfare that began in Ukraine is not merely a tactical or operational shift. It represents a fundamental change in the relationship between industrial capacity, economic resources, and military power. The lesson is clear: in future conflicts, the side that can produce more, faster, and cheaper will possess advantage regardless of the sophistication of individual systems. The economics of warfare have been permanently transformed.

As we look at the ongoing conflict, now in its fourth year, the economic dimension proves as decisive as the tactical or strategic. Ukraine survives not because it possesses superior technology, but because it built an industrial base capable of producing millions of drones annually. Russia maintains its offensive despite heavy losses because it too achieved mass production. The war continues because both sides can economically sustain the attrition.

The ultimate lesson for military planners, political leaders, and citizens is uncomfortable but unavoidable: prepare not for the war you can afford with peacetime economics, but for the war that requires total industrial mobilization. The drone war has shown us that modern conflicts between committed opponents will be won not on battlefields alone, but in factories, workshops, and distributed manufacturing networks churning out weapons by the thousands. The side with deeper industrial capacity and more resilient production will prevail, regardless of how sophisticated their peacetime arsenal appeared.

This is the new arithmetic of warfare, written in the skies over Ukraine in drones numbered in the millions. The calculations are simple, brutal, and impossible to ignore.

Chapter 23

Counter-Drone Technologies

On the night of October 9, 2025, Ukraine's integrated air defense system faced one of its most severe tests of the war. Russian forces launched over 400 drones into Ukrainian airspace in a coordinated mass attack designed to overwhelm defenses through sheer saturation. Air raid sirens wailed across every major city. Millions of Ukrainians descended into shelters, many carrying the small emergency bags that had become standard household items. Children clutched toys and tablets, their faces showing the weary resignation of those who had lived through hundreds of such nights.

What happened next represented both a triumph of defensive technology and a sobering reminder of the challenges inherent in stopping drone swarms. Ukrainian air defenses destroyed 93% of the attacking drones, an achievement that would have seemed impossible just two years earlier. Yet 28 drones penetrated the defensive net, striking targets across the country. Another 19 unarmed decoy drones crossed into Polish airspace, triggering NATO fighter scrambles and creating a diplomatic crisis that lasted for days. Four of those drones were destroyed over Poland by allied interceptors.

This single engagement encapsulates the central paradox of counter-drone warfare: even highly effective defenses will allow some attackers through when facing mass saturation, and the consequences of those penetrations can be severe. The challenge facing Ukraine, NATO, and military establishments worldwide is not merely to improve interception rates, but to fundamentally rethink how to defend against weapons that cost a tiny fraction of traditional defensive systems.

The Evolution of Ukraine's Air Defense

Ukraine began the war with a hodgepodge of Soviet-era systems, some dating to the 1970s and 1980s. S-300 missile batteries, designed to shoot down high-flying aircraft, formed the backbone of the country's defenses.

These systems were never intended to counter the low, slow, small targets that drones represent. Their radars struggled to detect small quadcopters. Their missiles, costing over $1 million each, were grotesquely expensive solutions to cheap threats.

The first months of the war exposed these limitations brutally. Russian drones flew reconnaissance missions with near impunity. When the Shahed-136 attacks began in August 2022, Ukraine initially had few effective responses. The foam and plywood construction of these kamikaze drones made them nearly invisible to radar systems optimized for detecting metal aircraft. They flew too low for many missile systems to engage effectively. Their slow speed, only about 185 kilometers per hour, paradoxically made them harder to intercept than faster targets.

Ukraine's defensive evolution proceeded along multiple parallel tracks, each addressing different aspects of the drone threat. The transformation was not orderly or centrally planned. Instead, it emerged from desperate innovation by frontline units, civilian volunteers, and military planners working with whatever resources they could acquire or improvise.

The first major breakthrough came with the integration of mobile anti-aircraft gun systems. Germany's provision of Gepard self-propelled anti-aircraft guns in summer 2022 provided Ukraine with platforms specifically designed for low-altitude threats. The Gepard's dual 35mm autocannons, controlled by sophisticated fire control radar, proved remarkably effective against drones. Each gun could fire 550 rounds per minute, creating a wall of steel that small drones struggled to penetrate.

The economics of the Gepard system were far more favorable than missiles, though still imperfect. At $30-50 per round, a typical engagement expending 100-200 rounds cost $3,000-10,000, expensive compared to the target, but manageable compared to million-dollar missiles. More importantly, the Gepard could engage multiple targets in quick succession, something missile systems struggled to do.

Other nations contributed their own systems, creating a diverse defensive ecosystem. Norwegian NASAMS (National Advanced Surface-to-Air Missile System) provided medium-range coverage with sophisticated radar integration. American Patriot batteries offered protection for critical targets, though their missiles remained prohibitively expensive for routine drone intercepts. British Starstreak systems, French Crotale batteries, and Polish Grom man-portable missiles all found roles in the expanding defensive network.

But the real transformation came through innovation in three critical areas: electronic warfare, integrated early warning networks, and the emergence of counter-drone drones.

Electronic Warfare: The Invisible Battle

Electronic warfare systems represent the most cost-effective counter to drones, at least in theory. A jamming system that can disrupt drone communications or GPS signals provides intercepts at the cost of electricity rather than munitions. For a nation facing thousands of drone attacks monthly, this economic advantage is decisive.

Ukraine's electronic warfare capabilities evolved rapidly under combat pressure. In early 2022, the country possessed some Soviet-era jamming equipment and a small number of more modern systems. By 2025, Ukraine had become arguably the world leader in tactical counter-drone electronic warfare, with systems ranging from man-portable jammers to vehicle-mounted stations capable of disrupting entire sectors of airspace.

The technology operates on several principles. The simplest systems broadcast interference on the radio frequencies used by commercial drones for control and video transmission. Most consumer drones operate on 2.4

GHz or 5.8 GHz bands, the same frequencies used by WiFi routers. Flooding these frequencies with noise disrupts the connection between operator and drone. In many cases, the drone's programming responds by returning to its launch point or landing immediately, a defensive victory without firing a shot.

GPS spoofing represents a more sophisticated approach. Rather than simply jamming navigation signals, spoofing systems broadcast false GPS data. A drone receiving this false information believes it is in the wrong location and corrects its course accordingly, often flying away from its intended target or even returning toward its own lines. Ukrainian operators reported numerous instances of Russian drones landing intact behind Ukrainian lines, presumably due to GPS spoofing that convinced them they were over friendly territory.

The cat-and-mouse dynamic in electronic warfare escalated throughout the conflict. As Ukraine deployed jammers more widely, Russia developed drones with anti-jam features: frequency hopping, encrypted command links, and autonomous navigation that didn't rely on GPS. Ukraine responded with more sophisticated jamming, broadband interference, and systems that could adapt in real-time to enemy countermeasures.

The introduction of fiber-optic cable guidance by Russia in spring 2024 represented a major challenge to electronic warfare approaches. A drone trailing a fiber-optic cable behind it is immune to radio jamming and GPS spoofing where the physical cable provides a jam-proof connection. Ukraine rapidly adopted the same technology, creating a segment of the drone war where electronic warfare provided no advantage. Both sides then focused on developing hybrid systems: fiber-optic for the terminal approach to the target, but radio or autonomous guidance for the initial flight phase.

Colonel Yuriy Fedorenko, commander of Ukraine's 92nd Mechanized Brigade's electronic warfare detachment, explained the complexity in an interview from October 2024: "Every time we develop a solution, they adapt. Every time they adapt, we find a new solution. It is evolving at speed. A system that works today may be obsolete in two weeks. We must constantly innovate or lose the advantage."

The effectiveness of electronic warfare varies enormously by environment and circumstances. In the open terrain of southern Ukraine, long-range jamming can create protective bubbles dozens of kilometers across. In urban environments, buildings reflect and absorb radio signals, creating shadows where drones can operate. Against older commercial drones, jamming success rates approach 80-90%. Against newer military drones with anti-jam capabilities, rates drop to 30-40%. Against fiber-optic guided drones in their terminal phase, electronic warfare provides no defense at all.

The Network Approach

Ukraine's success in achieving 90-93% interception rates against mass drone attacks stems not from any single technology, but from sophisticated integration of multiple systems into a cohesive network. This integrated approach combines early warning, threat assessment, optimal resource allocation, and layered defenses in ways that maximize effectiveness while minimizing cost.

The foundation is early warning. Ukraine developed a distributed network of observers, acoustic sensors, and radars providing real-time detection of inbound drones. Mobile applications allow civilians to report drone sightings, feeding data into centralized systems that track threats as they develop. This crowdsourced intelligence proved remarkably effective; a drone spotted by a citizen 100 kilometers from its target provides warning time that allows defensive resources to be positioned optimally.

Acoustic detection systems emerged as a critical capability. Drones produce distinctive sounds, the buzz of small quadcopters, the lawn-mower-like drone of Shaheds. Microphone arrays can detect and triangulate these sounds, providing detection when radar fails. Ukrainian companies developed specialized acoustic sensors deployed by the hundreds around critical targets. Algorithms learned to distinguish drone sounds from aircraft, vehicles, and background noise, achieving detection ranges of several kilometers for larger drones.

The integration of all these sensors into a common operating picture represents one of Ukraine's most significant technical achievements. The system, developed through a combination of Western technology, Ukrainian software engineering, and combat experience, provides air defense commanders with real-time awareness of all detected threats. Artificial intelligence algorithms predict drone trajectories, assess threat priorities, and recommend optimal engagement solutions.

This allows Ukraine to apply the principle of "right tool for right threat." Expensive missiles are reserved for high-value targets or drones penetrating close to critical infrastructure. Gun systems engage medium-priority threats at medium range. Electronic warfare handles low-priority targets or ambiguous contacts. Mobile units reposition dynamically based on evolving threats. The result is a defense that operates far more efficiently than the sum of its parts.

The October 9 engagement demonstrated both the strengths and limitations of this approach. The 93% interception rate reflected superb integration and execution. But 400 attacking drones overwhelmed the system through sheer mass. Some drones saturated defenses in specific sectors. Others exploited gaps in coverage. A few simply got lucky, passing through defended zones without being engaged. The 28 that penetrated caused significant damage, demonstrating that even a highly effective defense is not an impenetrable shield.

Counter-Drone Drones

Perhaps the most conceptually elegant solution to the drone problem is to use drones to kill drones. The economics are favorable, a $2,000 interceptor drone engaging a $500 attack drone represents a sustainable exchange ratio. The technical matchup is appropriate, drones can pursue other drones through complex environments where missiles and guns struggle. And psychologically, there is satisfaction in turning the enemy's own paradigm against them.

Ukraine began experimenting with counter-drone drones in late 2023. The early attempts were crude: FPV drones flown manually to intercept incoming threats. Success rates were low and the approach required skilled operators. But the concept proved sound enough to warrant continued development.

By 2024, specialized interceptor drones began appearing. These platforms differed from attack drones in several key ways. They prioritized speed and maneuverability over payload capacity. They incorporated sophisticated sensors, cameras with zoom capability, thermal imaging, and in some cases basic radar. Most importantly, they included increasing levels of automation to reduce operator workload.

The most advanced systems operated on a "human-supervised autonomy" model. An operator would designate a target on their screen, an incoming drone detected by the air defense network. The interceptor drone would autonomously fly to intercept, maneuvering as needed to close with the target. At close range, either the operator would take manual control for the final engagement, or the system would autonomously ram or detonate near the target.

These interceptor drones proved particularly effective against Shahed-136 attacks. The Shaheds' relatively slow speed and predictable flight paths made them vulnerable to interception. Ukrainian interceptors would patrol predicted approach routes, waiting for targets to appear. Once detected, the interceptor would accelerate to overtake the Shahed from behind. Shaheds have no rearward sensors and either ram it or detonate a small explosive charge. Success rates for these engagements reached 60-70% under optimal conditions.

Against faster, more maneuverable FPV drones, interception proved more challenging. Both drones operate at similar speeds and both are highly agile. The engagements resembled aerial dogfights in miniature, with both operators maneuvering frantically. Success often depended on which operator had better situational awareness, quicker reflexes, or simply got lucky. These drone-versus-drone duels, captured on video and shared widely on social media, represented a genuinely new form of aerial combat.

The limitations of counter-drone drones became apparent through operational experience. They required skilled operators, though less skilled than attack drone pilots. They struggled in bad weather when visibility limited engagement ranges. They had limited endurance, typically 15-30 minutes requiring many interceptors on station to maintain continuous coverage. And they remained vulnerable to the same electronic warfare that affected attack drones.

Despite these limitations, counter-drone drones became an increasingly important part of Ukraine's defensive architecture. By late 2025, specialized interceptor units operated around major cities and critical infrastructure, providing a middle layer between electronic warfare and expensive missiles. The units developed sophisticated tactics: patrol patterns that maximized coverage, coordinated intercepts where multiple defenders engaged single high-value targets, and rapid response protocols that got interceptors airborne within seconds of threat detection.

Directed Energy: The Future Arriving

Throughout 2024 and into 2025, military planners worldwide focused increasing attention on directed energy weapons, lasers and high-powered microwaves as potential game-changers in counter-drone warfare. The theoretical advantages are compelling. A laser system fires shots that travel at the speed of light, cost pennies in electricity, and require no ammunition logistics. A high-powered microwave can disable multiple drones simultaneously, frying their electronics. These capabilities promise to solve the economic challenge that makes drone defense so difficult.

Ukraine, with support from Western partners, began fielding prototype directed energy systems in late 2024. The initial deployments were limited and experimental, but they provided valuable operational data on what these technologies could and could not do.

Laser systems proved effective against drones under specific conditions. A focused beam delivering 10-50 kilowatts of energy could burn through a drone's plastic housing in seconds, destroying control electronics or igniting fuel or warheads. Against hovering or slow-moving drones at ranges under two kilometers, success rates approached 80-90%. The engagement cost was negligible, a few dollars of electricity per intercept.

But lasers struggled with several fundamental limitations. Weather severely degraded performance. Clouds, fog, rain, or even high humidity scattered and absorbed laser energy, dramatically reducing effective range. At night, lasers worked well. During daytime, atmospheric effects increased. Against fast-moving targets executing evasive maneuvers, the laser's requirement to hold the beam on target for several seconds became problematic. And the systems themselves remained expensive, several million dollars per installation and required substantial electrical power generation.

High-powered microwave systems attacked the problem differently. Instead of focusing energy on a single point, they radiated broad areas with electromagnetic energy intense enough to induce electrical failures in drone systems. The theoretical advantage was engaging multiple targets simultaneously and being less weather-dependent than lasers.

Operational testing revealed mixed results. Against older commercial drones with minimal shielding, HPM systems proved devastatingly effective, disabling everything within their engagement zone. Against newer military drones with electromagnetic hardening, effectiveness dropped dramatically. The systems also suffered from limited range, typically under one kilometer and posed risks to friendly electronics within the engagement area.

By late 2025, directed energy weapons remained in the experimental phase. They showed enormous promise but were not yet mature enough for widespread deployment. Most experts predicted that by the late 2020s, these technologies would become standard components of integrated air defenses. For now, they represented the future arriving slowly, offering tantalizing glimpses of post-missile air defense but not yet delivering on their full potential.

Low-Altitude Air Defense Gaps

Even with all these technologies, Ukraine's counter-drone experience revealed a persistent problem that no current system adequately solves: the low-altitude gap. Drones operating below 100 meters altitude, especially in urban or forested terrain, remain extraordinarily difficult to detect and engage.

Traditional air defense radars are designed to search the skies for aircraft at medium to high altitudes. Their radar beams, aimed upward, create a cone of coverage that leaves low-altitude zones poorly protected. Ground clutter, reflections from buildings, trees, and terrain creates false contacts that overwhelm radar processing. A small drone flying 50 meters above treetops presents a radar signature similar to birds, vehicles, or wind-blown debris.

Optical and acoustic sensors help but have limited range. A camera can only see a few kilometers even in clear weather. Acoustic sensors struggle to pinpoint drones over background urban noise. Electronic warfare provides no defense against autonomous or fiber-optic guided drones. Interceptor drones must know where to look, requiring other sensors to provide targeting data.

The result is a persistent vulnerability that both sides exploited throughout the conflict. Russian quadcopter drones regularly reached deep into Ukrainian-controlled cities, flying below the radar horizon and evading defensive systems. These drones dropped internationally prohibited landmines in Kherson, attacked civilian gatherings in markets, and struck rear-area military targets. Ukrainian forces operating in Russia during the Kursk offensive faced similar threats from Russian drones that operated in the low-altitude gap with relative impunity.

Solving the low-altitude problem requires fundamentally different approaches. Some nations experimented with dense networks of short-range sensors, essentially putting detection capabilities every few hundred meters. Others developed mobile gun systems optimized for low-altitude engagements. AI-enabled acoustic networks showed promise by providing three-dimensional tracking of multiple low-altitude threats simultaneously.

But as of late 2025, no complete solution existed. The low-altitude gap remained exactly that, a gap in coverage that ensured determined attackers could penetrate defenses if they were willing to accept higher losses. This reality forced a sobering conclusion: there is no perfect defense against drone swarms. Highly effective, yes. Adequate for most purposes, certainly. But impenetrable? No.

NATO Learning from Ukrainian Experience

The counter-drone innovations emerging from Ukraine did not go unnoticed by NATO militaries. Alliance members recognized that Ukrainian forces were writing the playbook for drone defense through painful trial and error. What Ukraine learned at the cost of thousands of civilian casualties and billions in damaged infrastructure, NATO could learn by observation and adaptation.

Multiple NATO countries established study teams in Ukraine, examining defensive systems, interviewing operators, and documenting tactics and procedures. The lessons were sobering. Most NATO militaries had minimal dedicated counter-drone capabilities beyond individual systems for base defense. Their air defense architectures, optimized for Cold War threats like bombers and cruise missiles, were poorly suited for mass drone attacks.

NATO response proceeded along several lines. Nations accelerated procurement of systems proven effective in Ukraine: additional Gepard guns, mobile short-range air defense systems, electronic warfare equipment, and interceptor drones. Military exercises increasingly incorporated mass drone attack scenarios. Doctrine writers worked to codify Ukrainian experience into formal counter-drone tactics, techniques, and procedures.

Perhaps most significantly, NATO began rethinking air defense architecture from the ground up. The traditional model of a few expensive, highly capable systems protecting vast areas clearly could not handle saturation attacks by cheap drones. The emerging model instead emphasized layered defenses with much greater redundancy: dense sensor networks, multiple overlapping engagement systems at different altitudes and ranges, electronic warfare as a first line of defense, and kinetic systems in depth.

The October 9, 2025 incident, with Russian drones entering Polish airspace and being destroyed by NATO fighters, provided a sharp reminder that the drone problem extended beyond Ukraine's borders. If Russia was willing to send drones into NATO territory, even if unintentionally then alliance air defenses needed capabilities they currently lacked.

The Persistent Challenge

As Ukraine's war entered its fourth year, the counter-drone mission remained one of constant adaptation and evolution. Every defensive innovation prompted offensive counter-innovation. Every successful engagement type led to new attack profiles designed to exploit defensive weaknesses. The cycle never ended.

Ukrainian air defenders achieved remarkable success by any reasonable measure. Interception rates that would have been impossible to imagine in 2022 became routine by 2025. But "routine" meant that on a typical night, dozens of drones still penetrated defenses. Civilians still died. Infrastructure still burned. The economic burden of defense still mounted.

The fundamental challenge remained unchanged from that first moment when a Ukrainian commander watched $500 drones flying toward targets defended by million-dollar missiles: the economics favor the attacker. Until defensive costs drop to match offensive costs, the balance tilts toward whoever can produce and launch more drones. Ukraine achieved this through the comprehensive integration of multiple technologies, operational excellence, and constant innovation under fire.

The lesson for other nations is clear but daunting. Effective counter-drone defense requires not just technology, but entire ecosystems of capabilities working in concert. It requires sensors distributed across wide areas, operators trained to make split-second decisions, systems integrated to share information seamlessly, and doctrine that balances multiple engagement options against varying threats. Most critically, it requires recognition that perfect defense is impossible. The goal is not to stop every drone, but to make drone attacks costly enough and ineffective enough that they lose strategic utility.

That goal remains elusive. As long as drones can be produced for hundreds of dollars and defenses cost thousands or millions, attackers willing to accept losses can overwhelm even the most sophisticated defenses through sheer mass. The race to develop technologies that change this calculus such as directed energy, advanced autonomous systems, ultra-low-cost interceptors continues. Until those technologies mature, counter-drone defense remains an art rather than a science, practiced by operators who must win every engagement while their opponents need only win occasionally.

Ukraine's 93% success rate on that October night represented a triumph of human skill, technological integration, and organizational excellence. It also represented the limits of what current technology can achieve against mass drone warfare. Those limits, written in the skies over Ukraine in fire and falling debris, define the challenge facing every military in the world.

Chapter 24

The AI and Autonomy Frontier

In a nondescript office building in Kyiv, far from the front lines, a team of Ukrainian software engineers worked through the night in March 2025 on a problem that would have seemed like science fiction just three years earlier. Their challenge: teach a drone to recognize and engage targets without human intervention in GPS-denied environments. The stakes were existential. Russian electronic warfare had become so sophisticated that traditional drone guidance was increasingly ineffective. Ukraine needed drones that could think for themselves.

By dawn, they had a prototype working. A small quadcopter equipped with a camera and a modified smartphone processor could distinguish between civilian vehicles and military targets with 87% accuracy. It could navigate using visual landmarks when GPS was jammed. Most significantly, it could autonomously prosecute an attack from identification through impact without any human input beyond the initial launch command.

The engineers celebrated briefly, then returned to work. They knew the Russians were working on similar capabilities. They knew that what they had created raised profound ethical questions. And they knew that in warfare, the side that hesitates loses. The age of fully autonomous combat drones had arrived, born not in sterile research laboratories but in the desperate crucible of existential war.

The Path to Autonomy

The journey toward autonomous drones in Ukraine was not a planned progression but an evolution forced by necessity. In the war's early months, every drone required constant human control. Operators flew reconnaissance missions manually, watching video feeds and adjusting flight paths in real-time. Attack drones were piloted all the way to impact, with operators making split-second decisions about targeting and weapon release.

This model worked adequately when electronic warfare was primitive and drone operations were relatively small-scale. But as both sides deployed increasingly sophisticated jamming, GPS spoofing, and signal disruption, purely manual control became untenable. Drones lost connection with operators at critical moments. GPS-guided systems flew wildly off course or circled uselessly when their navigation was compromised. Video feeds froze or displayed static during terminal attack phases.

The first response was semi-autonomous operation. Drones received waypoint navigation, fly to point A, then point B, then C that allowed them to continue missions even when communication was interrupted. If the connection dropped, the drone would complete its programmed route and return to base. This kept drones operational but limited flexibility. A target that moved after mission planning could not be engaged. Unexpected threats or obstacles often resulted in mission failure or drone loss.

The breakthrough came with the integration of machine vision and artificial intelligence. Using commercially available machine learning frameworks, many developed by major tech companies for completely different purposes, Ukrainian developers created systems that could process camera feeds in real-time and make decisions based on what they saw.

Early implementations were crude. A system trained on thousands of images of Russian tanks could identify tanks in drone camera footage with reasonable accuracy. The drone could then autonomously steer toward that tank, even without GPS or operator commands. The accuracy was imperfect, the system sometimes mistook civilian vehicles for military targets, or failed to recognize targets in challenging lighting or weather. But it was a start.

Dmytro Kovalenko, an AI researcher who transitioned from working on facial recognition software to military applications in 2023, described the rapid progression: "We compressed what should have been five years of development into six months. Every week brought new training data from the front. Every failure was analyzed immediately. We had something no peacetime research lab ever had, instant and brutal feedback on what works and what doesn't."

By mid-2024, Ukrainian autonomous systems had advanced dramatically. Drones could distinguish between different types of military vehicles. They could track moving targets, adjusting their attack vectors to intercept. They could navigate complex urban environments using visual navigation, recognizing landmarks and following roads without GPS. Most controversially, they could select and engage targets based entirely on algorithmic decisions.

Russia pursued parallel development, with some evidence suggesting they achieved certain capabilities before Ukraine. Russian reconnaissance drones demonstrated sophisticated autonomous tracking features by early 2024. Their Lancet loitering munitions showed increasing autonomy in target selection and engagement. By late 2024, both sides operated drones with significant autonomous capabilities, though the exact extent remained closely guarded.

Machine Vision and Target Recognition

The core enabling technology for autonomous combat drones is machine vision, the ability of computers to "see" and interpret visual information. This technology, developed primarily for civilian applications like autonomous vehicles and industrial robotics, found immediate military application in Ukraine.

The fundamental challenge is one of pattern recognition. A human observer can instantly distinguish a tank from a bus, a soldier from a civilian, an ammunition dump from a warehouse. Teaching computers to make these same distinctions requires training them on enormous datasets of labeled images. The system learns patterns: tanks have specific shapes, treads, turrets. Military vehicles typically appear in certain contexts, on roads near front lines, in defensive positions and camouflaged in tree lines.

Ukrainian developers had access to unprecedented quantities of training data. Every reconnaissance flight captured thousands of images. Drone footage from attacks provided clear examples of valid targets. Soldiers at the front used smartphone apps to photograph and categorize Russian equipment. Within months, Ukraine had amassed one of the world's largest datasets of military equipment imagery, all labeled and categorized.

Training neural networks on this data produced systems with remarkable capabilities. By late 2024, Ukrainian target recognition algorithms could identify specific vehicle types, not just "tank" but "T-72B3" versus "T-80BVM." They could recognize military installations, distinguish between occupied and abandoned positions, and even identify individuals carrying weapons versus those who were unarmed.

The systems worked by analyzing multiple features simultaneously. Shape and size provided initial classification. Thermal signatures helped distinguish active vehicles from derelict ones. Movement patterns indicated military versus civilian activity. Context mattered, a truck near a front-line trench was likely military; the same truck in a city center probably civilian. The algorithms assigned probability scores: 95% confidence this is a military target, 60% confidence, 30% confidence.

But machine vision had significant limitations. Camouflage and concealment specifically designed to fool computer vision could reduce recognition rates dramatically. Bad weather, darkness, or partial obscurement created ambiguity. The algorithms sometimes exhibited unpredictable behavior, making confident assessments on poor data or missing obvious targets that appeared unusual. And they were vulnerable to adversarial techniques, patterns designed to confuse neural networks, causing them to misclassify images in predictable ways.

Perhaps most concerning was the question of error rates. A human operator who correctly identifies 95% of targets is considered excellent. But that same 5% error rate, applied to hundreds of autonomous drone strikes daily, meant dozens of potential misidentifications. When those errors involved striking civilian targets instead of military ones, the consequences were catastrophic.

Autonomous Navigation in GPS-Denied Environments

GPS jamming and spoofing created zones where satellite navigation became unreliable or completely unavailable. For drones dependent on GPS for navigation, these zones became dead spaces where operations were impossible. Autonomous navigation systems that didn't rely on GPS became critical capabilities.

The solution drew on techniques developed for indoor robotics and autonomous vehicles. Visual-inertial navigation combined camera feeds with accelerometers and gyroscopes to track position and orientation. The drone would analyze its camera feed, identifying distinctive landmarks and tracking how its position relative to those landmarks changed. Internal sensors measured acceleration and rotation. Combining these inputs, the system could maintain a reasonably accurate position estimate even without GPS.

Ukrainian developers implemented visual navigation in stages. Early systems required pre-mapped routes. The drone would be shown photographs of landmarks along its route, a distinctive building, an intersection, and a damaged bridge. During flight, it would match what its camera saw to these reference images, confirming it was on course. If it deviated, it would correct back toward the planned route.

More sophisticated systems used simultaneous localization and mapping, SLAM in robotics terminology. The drone would build a three-dimensional map of its environment as it flew, simultaneously tracking its position within that map. This allowed operation in completely unknown areas without any pre-mapping. The drone could navigate based purely on what it observed in real-time.

These systems enabled operations that were previously impossible. Fiber-optic guided drones could use autonomous navigation for their initial flight phase, only deploying the cable as they approached their targets. This extended their effective range dramatically while maintaining jam-resistance in the terminal phase. Long-range strike drones could fly hundreds of kilometers through heavily jammed areas, navigating by visual landmarks and terrain features.

The limitations were significant. Visual navigation required good visibility, clouds, fog, darkness, or smoke could degrade or eliminate capability. Complex environments with few distinctive landmarks made position estimation difficult. And the computational requirements were substantial, requiring relatively powerful processors that added cost, weight, and power consumption.

Despite these challenges, autonomous navigation became standard in Ukrainian drones by late 2024. The capability to operate in GPS-denied environments was no longer a luxury but a necessity. Russian electronic warfare had ensured that any drone dependent solely on satellite navigation was obsolete.

Swarming Technologies

Perhaps no aspect of autonomous drone warfare captured public imagination more than the concept of drone swarms, large numbers of drones operating in coordination, attacking from multiple directions simultaneously, overwhelming defenses through sheer numbers and coordinated tactics. The reality of swarming technology in the Ukraine war fell short of science fiction depictions, but real capabilities emerged nonetheless.

True swarming requires drones to communicate with each other, coordinate their actions, and adapt to changing situations collectively. Early Ukrainian experiments with coordinated attacks used simple pre-programmed timing. Multiple drones would launch at carefully calculated intervals, scheduled to arrive at the target simultaneously from different directions. This wasn't true swarming, if one drone was intercepted or delayed, the others continued with their original plans but it achieved the basic goal of simultaneous multi-axis attack.

More sophisticated approaches emerged through 2024. Drones equipped with radio links could share basic information: "I've been intercepted," "I've found the target," "Defensive fire from this location." Other drones in the group could adjust their tactics accordingly. If the lead drone was shot down, the second drone would know to approach from a different direction. If one drone identified the target, others could converge on that location without needing to search independently.

The technical challenges were substantial. Radio communication between drones could be jammed just like operator control links. Drones had limited computational power for processing information from other members of the swarm. Coordinating tactics required sophisticated algorithms that could adapt to rapidly changing situations. And every additional capability added weight, cost, and complexity.

By late 2025, both Ukrainian and Russian forces demonstrated limited swarming capabilities. Groups of 5-10 drones could coordinate attacks with basic communication and adaptation. Claims of larger swarms operating autonomously remained unverified and were likely exaggerated.

The most effective "swarms" were actually large numbers of independently operating drones launched in coordinated waves, achieving similar effects without true swarming technology.

Research continued into more advanced swarming. Ukrainian developers explored hierarchical structures where one "leader" drone with more sophisticated electronics coordinated multiple "follower" drones with minimal intelligence. Others investigated mesh networks allowing any drone to communicate with any other, creating resilient command structures that survived the loss of individual members. Still others worked on emergent behavior models where simple rules followed by all drones produced complex collective tactics.

The military implications of mature swarming technology were profound. A swarm of 100 drones attacking a single target could saturate even sophisticated defenses. Swarms could conduct reconnaissance across vast areas simultaneously, compressing what would take hours into minutes. Adaptive swarms could respond to enemy countermeasures in real-time, making defensive planning nearly impossible.

But as of late 2025, these remained largely future capabilities. The swarms flying over Ukraine were more coordinated than truly intelligent, more simultaneous than adaptive. The path toward science-fiction-level swarming was clear, but significant technical hurdles remained.

The Human Control Question

As drones became increasingly autonomous, a fundamental question emerged: how much human control should remain over lethal decisions? This was not merely an ethical or legal question but a practical tactical one. Too much human oversight limited the speed and effectiveness of autonomous systems. Too little risked catastrophic mistakes or escalation beyond human intent.

International humanitarian law requires human control over targeting decisions, the principle of "meaningful human control" over weapons. But what constitutes "meaningful" control remained undefined and contested. A human operator who designates a target area and launches a drone that autonomously engages anything within that area, is this meaningful control? What about a system that identifies targets and asks for human approval before firing? Or one that engages autonomously but allows a human to abort the attack at any time?

Ukraine grappled with these questions under combat pressure. The initial approach required human confirmation for every engagement. An autonomous system would identify and track a target, but an operator would review camera footage and authorize the attack. This maximized human oversight but created bottlenecks. Operators struggled to review multiple targets simultaneously. Delays allowed targets to escape or defenses to respond.

By mid-2024, many Ukrainian systems operated with different authorization levels based on confidence scores. High-confidence targets in designated military zones could be engaged autonomously. Lower-confidence targets required human review. Targets in areas where civilians might be present always required human authorization. This hybrid approach balanced effectiveness with ethical constraints.

Russia appeared to adopt more permissive autonomous engagement rules, based on analysis of attack patterns and Ukrainian reports of drone behavior. Russian drones increasingly struck targets with minimal evidence of human guidance in the terminal phase. Some attacks on civilian areas showed patterns consistent with autonomous target selection based on algorithmic priorities rather than human judgment.

The practical reality was that increasing autonomy was militarily necessary. Electronic warfare degraded or eliminated human control over drones during critical attack phases. Time-sensitive targets required immediate engagement before they could escape or hide. Human operators suffered from fatigue, stress, and limited attention spans when controlling multiple drones simultaneously. Autonomous systems never tired, never hesitated, never got emotionally overwhelmed.

Lieutenant Oleksandr Ivanov, a drone company commander who transitioned from manual control to supervising autonomous systems, described the evolution: "In 2022, I flew every mission myself. By 2024, I was supervising eight drones simultaneously, authorizing autonomous engagements, intervening only when the system was uncertain. By 2025, my role became more like an air traffic controller than a pilot. I set objectives and constraints, the systems executed. I could override any decision, but usually I didn't need to."

This shift toward supervisory control represented a fundamental change in the human-machine relationship in warfare. Humans remained in the loop, but as managers and overseers rather than direct operators. The immediate tactical decisions of when to strike, which approach vector to use and how to react to defensive fire increasingly fell to algorithms.

Ethical and Legal Minefields

The development of increasingly autonomous weapons in Ukraine occurred outside the normal frameworks of international arms control and humanitarian law. Neither Ukraine nor Russia were parties to ongoing discussions about lethal autonomous weapons systems at the UN Convention on Certain Conventional Weapons. The urgent necessities of war overrode concerns about establishing precedents or violating emerging norms.

This created a legal and ethical minefield. International humanitarian law requires that attacks distinguish between combatants and civilians, be proportionate to military advantage, and avoid unnecessary suffering. Can an algorithm make these distinctions reliably? Does the commander who launches an autonomous drone bear responsibility for everything it does? What about the programmer who wrote the target recognition software?

Documented incidents raised troubling questions. In several cases, Ukrainian autonomous drones struck targets that appeared military from certain angles but were civilian vehicles. Russian autonomous systems attacked civilian infrastructure with patterns suggesting algorithmic target selection. Both sides deployed landmine-dropping drones operating autonomously, weapons that violated international treaties but were highly effective tactically.

Human rights organizations documented these incidents and expressed alarm at the erosion of human control over lethal force. Yet their criticisms gained little traction in Ukraine, where the alternative to autonomous drones was defeat. Ethical principles that seemed paramount in peacetime felt like luxuries a nation fighting for survival could not afford.

Ukrainian officials publicly maintained that all lethal engagements involved meaningful human control. In practice, "meaningful" stretched to cover systems where human intervention was possible but rare, where operator oversight was nominal, where algorithms made the real targeting decisions. This gap between official position and operational reality reflected the impossible situation Ukraine faced: acknowledging full autonomy would invite international condemnation, but fighting without it would risk military defeat.

The absence of international oversight or constraint meant each side pushed autonomy as far as their technology and risk tolerance allowed. No neutral observers verified claims about human control. No enforcement mechanisms existed even if violations were documented. The only constraints were technical limitations and whatever moral boundaries combatants chose to observe.

Professor Maria Shevchenko of Kyiv University, one of Ukraine's few ethicists still actively publishing during the war, wrote in October 2024: "We are establishing precedents that will shape warfare for generations, but we are doing so without debate, without consensus, without safeguards. Every autonomous kill normalizes the next. When this war ends, the world will be left with autonomous weapons proliferated globally and norms eroded irreparably. We must ask whether defeating our present enemy is worth arming every future one."

Her warnings went largely unheeded. The imperatives of survival overwhelmed abstract ethical concerns. The autonomous weapons developed in Ukraine would indeed spread globally, but stopping that proliferation would have required Ukraine to accept military disadvantages that might prove fatal.

The AI Arms Race

By late 2025, the conflict had evolved into an AI arms race where computational capabilities mattered as much as industrial production. Both sides poured resources into developing more sophisticated target recognition, better autonomous navigation, more effective swarming, and more resilient systems that could operate under all conditions.

Ukraine possessed advantages in software engineering talent and access to Western AI research. Ukrainian developers could leverage open-source machine learning frameworks, commercial cloud computing for training models, and partnerships with Western tech companies. Russia had its own significant computer science capabilities, less constrained by ethical concerns, and access to Chinese AI research and technologies.

The competition proceeded along multiple dimensions simultaneously. Better training data produced better target recognition. More efficient algorithms allowed more complex decision-making on limited drone processors. Improved sensor fusion combined cameras, thermal imaging, radar, and other inputs for more reliable target identification. Adversarial training created systems resistant to deception and spoofing.

Each advance prompted countermeasures. Better target recognition led to improved camouflage designed to fool algorithms. Autonomous navigation drove development of visual jamming such as smoke, lasers, or projected patterns that confused computer vision. Swarming capabilities encouraged counter-swarm tactics and weapons. The cycle of innovation and counter-innovation accelerated to speeds impossible with traditional weapons development.

The pace was unsustainable but neither side could afford to slow down. A Ukrainian autonomous system that was state-of-art in March might be obsolete by June, countered by Russian defensive measures. Software updates pushed to drones in the field weekly or even daily became routine. Operators described scenarios where their drones received updated AI models mid-mission, downloaded over radio links when they returned to communication range.

This software-defined warfare represented something genuinely new. Throughout military history, weapons remained relatively static once deployed. A rifle or tank produced in 1943 would have essentially identical capabilities in 1945. But a drone manufactured in January 2025 might have fundamentally different capabilities by March, transformed by software updates that rewrote its artificial intelligence.

The implications extended far beyond Ukraine. Every nation with significant AI capabilities watched the competition closely, learning what worked and what didn't. Chinese military planners studied Ukrainian autonomous systems intensively. American defense researchers analyzed Russian counter-drone AI. The techniques developed in Ukraine would spread globally within years if not months.

The Future Arriving Too Fast

As 2025 drew toward its close, autonomous drones had become routine in Ukraine. What would have seemed like science fiction in early 2022 was now standard operating procedure. Drones that selected and engaged targets without human input flew hundreds of missions daily. Systems that were navigated by artificial intelligence crossed vast distances through jammed environments. Swarms of primitive but functional coordinated attacks that no human could orchestrate.

The transformation happened so quickly that neither military doctrine nor international law kept pace. Officers commanded systems they barely understood, following procedures that hadn't existed months earlier. Lawyers struggled to apply laws written for human soldiers to algorithmic weapons. Ethicists debated questions that operators had already answered through necessity.

Yet for all the sophistication, limitations remained stark. Autonomous systems made mistakes that killed civilians. They could be fooled, jammed, or crashed. They struggled in conditions such as adverse weather, lighting and complex terrain where humans still excelled. The dream of fully autonomous warfare remained distant, replaced by the messy reality of human-machine teaming where neither worked optimally without the other.

The engineers who developed these systems understood both their capabilities and limitations with unusual clarity. They had no illusions about creating perfect weapons. They sought only to create weapons effective enough, reliable enough, survivable enough to keep Ukraine fighting. Whether they succeeded in building ethical autonomous systems remained an open question. That they succeeded in building operationally effective ones was undeniable.

The artificial intelligence revolution in warfare, predicted for decades by futurists and feared by ethicists, arrived not through careful planning but through desperate improvisation. It emerged not from defense laboratories but from civilian software developers repurposing commercial technology. And it happened not in some distant future but in real-time, written in code and tested in combat, on the battlefields of Ukraine between 2022 and 2025.

The genie was out of the bottle. Autonomous lethal systems were now proven combat capabilities, not theoretical concepts. The knowledge of how to build them had spread globally. The precedents for using them were established. The next war, wherever it occurred, would feature autonomous weapons more sophisticated than what Ukraine deployed. The war after that would push autonomy further still.

Ukraine had not chosen to become the laboratory for autonomous warfare. But history had chosen Ukraine, and Ukrainian engineers, soldiers, and civilians bore the burden of pioneering this technology under fire. The world would live with the consequences for decades to come.

Chapter 25

Strategic Implications

The transformation of drone warfare in Ukraine between 2022 and 2025 represents far more than a tactical evolution on a single battlefield. It marks a fundamental shift in the nature of modern conflict that will reverberate through military planning, defense budgets, and strategic calculations for decades to come. The lessons emerging from Ukraine's skies are being studied in war colleges from Beijing to Brussels, in Pentagon planning rooms and in the bunkers beneath the Kremlin. What began as improvised adaptation has become a blueprint for future warfare, one that challenges decades of military doctrine and threatens to upend the carefully balanced deterrence structures that have preserved an uneasy peace among great powers.

The Collapse of Strategic Depth

Perhaps the most profound implication of Ukraine's drone war is the effective collapse of strategic depth as a defensive concept. Throughout military history, distance has provided security. Armies could marshal forces behind front lines, regroup in rear areas, and maintain logistics hubs beyond the reach of enemy fire. Strategic bombers and ballistic missiles complicated this calculus in the twentieth century, but the expense and sophistication required to field such weapons meant they remained the province of major powers, deployed sparingly against the highest-value targets.

Drones have shattered this paradigm entirely. When Ukraine can strike oil refineries a thousand kilometers inside Russia with platforms costing tens of thousands of dollars, and Russia can send fiber-optic guided FPV drones into Ukrainian cities hundreds of kilometers from the front line, nowhere is truly safe. The "rear area" has ceased to exist in any meaningful sense. Supply depots, training facilities, command posts, and civilian infrastructure that once would have been secure simply by virtue of distance are now perpetually vulnerable.

This collapse has immediate implications for how nations must defend themselves. The traditional model of forward defense, concentrated at borders or along defensive lines, becomes inadequate when threats can materialize anywhere within national territory. Air defense networks, historically designed to protect against aircraft and missiles approaching from predictable vectors, must now account for low-altitude, small-signature threats that can emerge from any direction. The cost of providing comprehensive air defense coverage over an entire nation, rather than just key installations and frontline units, is staggering and perhaps impossible with traditional systems.

For smaller nations, particularly those lacking the industrial capacity to produce drones at scale or the resources to deploy comprehensive counter-drone defenses, this shift is existential. A neighbor with even a modest drone manufacturing capability can hold hostage critical infrastructure, economic centers, and population centers. Strategic depth, once measured in kilometers and mountain ranges, now means little. This reality will reshape alliances, drive military acquisitions, and influence everything from urban planning to industrial dispersal.

Deterrence in the Age of Cheap Precision Strike

The nuclear age gave birth to the concept of mutually assured destruction, a grim but stable form of deterrence predicated on the catastrophic consequences of great power conflict. Conventional deterrence operated on different principles: the demonstrated ability to defeat an adversary's military forces, the economic cost of conflict, and the political will to sustain casualties and expenditure. Both models assumed that war, if it came, would be enormously expensive in blood and treasure, creating powerful incentives for restraint.

Drone warfare introduces dangerous instability into these calculations. When a nation can conduct sustained strategic attacks using weapons that cost a fraction of what defenders must expend to stop them, the traditional cost-benefit analysis of conflict shifts dramatically. The economic deterrent weakens when attacks become affordable and defense becomes ruinously expensive. The threshold for initiating hostilities lowers when casualties can be inflicted remotely, without placing soldiers at risk, and when the financial barrier to sustaining offensive operations drops precipitously.

Russia's campaign against Ukrainian infrastructure illustrates this dynamic perfectly. Throughout 2024 and into 2025, Russia has launched thousands of Shahed drones against Ukrainian cities, power plants, and critical facilities. Each drone costs Russia approximately fifteen to thirty thousand dollars to produce. Ukraine, to defend against these attacks, must employ air defense missiles costing hundreds of thousands or millions of dollars each, deploy soldiers to manning positions, and maintain an extensive early warning and command network. Even with a ninety-three percent interception rate, the arithmetic is punishing. Russia can afford to launch hundreds of drones knowing that overwhelming Ukrainian defenses is a matter of production capacity rather than technological superiority.

This cost asymmetry creates new forms of coercion short of war. A nation facing economic pressure or political disputes might employ drone strikes as a form of graduated escalation, imposing costs without triggering the full-scale retaliation that would follow traditional military attacks. The line between peace and war blurs when strategic attacks can be conducted nightly without formal declarations, without mobilization, and without the diplomatic crisis that would attend bomber raids or missile strikes.

For NATO and the Western alliance, this poses uncomfortable questions about Article 5 commitments and collective defense. If a member state faces sustained drone attacks from a rival power, at what point does this constitute an armed attack triggering mutual defense obligations? The October 2025 incident in which Russian drones entered Polish airspace offered a glimpse of this dilemma. Poland scrambled fighters and destroyed four drones over its territory, but the question of whether repeated violations constitute an act of war requiring collective response remains unresolved. NATO's deterrent power rests on the credibility of its security guarantees, but drone warfare's ambiguous nature creates space for adversaries to probe and test those guarantees in ways that conventional forces could not.

Implications for NATO and European Defense

The NATO alliance enters the post-Ukraine period facing a strategic reckoning. For decades, NATO military planning centered on high-intensity conventional conflict, with emphasis on armored maneuver warfare, air superiority, and precision strike capabilities delivered by sophisticated aircraft and missile systems. The alliance invested trillions in fighter jets, main battle tanks, naval platforms, and the supporting infrastructure for traditional combined arms operations. This posture assumed that technological superiority, professional armed forces, and the industrial might of the United States and its allies would provide decisive advantages in any conflict with peer or near-peer adversaries.

Ukraine has demonstrated that this model is incomplete at best and potentially obsolete at worst. The conflict has shown that mass matters, that industrial capacity is a weapon system in itself, and that cheap, numerous platforms can overwhelm expensive, sophisticated defenses. NATO members watched Ukrainian forces armed with commercial drones and improvised systems battle Russian armored columns to a standstill in the war's opening months. They observed as both sides scaled production to millions of drones annually, creating an entirely new category of weapons expenditure that had barely existed in NATO budgets five years earlier.

The alliance now faces difficult choices about resource allocation. Should European nations reduce orders for F-35 fighter jets to fund crash programs for drone production? How much of traditional air defense budgets should shift toward counter-drone systems and electronic warfare capabilities? What is the appropriate balance between maintaining legacy capabilities and investing in the new paradigm? These are not merely procurement questions but fundamental choices about the future character of warfare and the nature of deterrence in Europe.

Several NATO members have begun to adapt. Poland, sharing a border with both Ukraine and Russia, has announced massive drone acquisition programs and domestic production initiatives. The Baltic states, acutely aware of their vulnerability, are developing comprehensive counter-drone networks and seeking to build indigenous drone manufacturing capabilities despite their small populations and limited industrial bases. Germany has pledged billions for military modernization with significant emphasis on unmanned systems. The United Kingdom is reorganizing military units to integrate drone warfare at every level of operations.

Yet adaptation is uneven and challenged by structural factors. NATO procurement systems, designed for multi-year development cycles, rigorous testing, and comprehensive oversight, move at a glacial pace compared to the rapid iteration happening in Ukrainian workshops and Russian factories. The alliance's emphasis on perfection, on zero-defect standards, and on gold-plated solutions clashes with the Ukrainian model of good-enough systems produced at scale. When TAF Drones can manufacture forty thousand units monthly in wartime conditions, NATO's traditional defense contractors, accustomed to delivering aircraft in the dozens over years, appear almost artisanal by comparison.

The geographic reality of European defense also complicates adaptation. NATO's eastern flank, from the Baltic Sea to the Black Sea, represents a frontier thousands of kilometers long. Providing comprehensive air defense coverage across this expanse against massed drone attacks would require an air defense architecture far more extensive and expensive than anything currently envisioned. The Ukraine experience suggests that even with such defenses, attrition would be severe and sustainability questionable. NATO must either accept greater risk, fundamentally rethink territorial defense concepts, or commit resources at a scale that may prove politically unsustainable in peacetime.

What China Is Learning

While Western analysts study Ukraine for insights into future NATO-Russia conflict scenarios, Chinese military planners focus on a different set of lessons. For Beijing, contemplating potential military operations against Taiwan, the drone war in Ukraine offers both opportunities and warnings that will shape People's Liberation Army doctrine and procurement for years to come.

The most obvious lesson concerns the challenge of amphibious operations under constant drone surveillance and attack. Taiwan Strait crossings, already among the most difficult military operations imaginable, become exponentially more dangerous in an environment where every ship, every landing craft, and every vehicle moving toward shore can be tracked and targeted by swarms of cheap, expendable drones. Ukraine has demonstrated that even well-equipped, modern militaries struggle to maintain operational security and force protection when drones saturate the battlefield. For an amphibious assault requiring precision timing, surprise, and the concentration of forces, the implications are sobering.

Chinese planners have undoubtedly noted how Ukrainian naval drones forced Russia's Black Sea Fleet to abandon Sevastopol and withdraw to more distant bases. They have studied how Turkish Bayraktar drones destroyed Russian armor in the war's opening weeks. They understand that Taiwan, an island with advanced industrial capacity and strong ties to Western technology suppliers, could rapidly scale drone production and potentially source systems from allies. The prospect of facing tens or hundreds of thousands of defensive drones while attempting to establish beachheads and supply lines across the Taiwan Strait introduces uncertainty that Chinese military models must now account for.

At the same time, China sees opportunities. The country is the world's dominant manufacturer of commercial drones and electronic components, the very technologies that both Ukraine and Russia have repurposed for military use. China's industrial capacity dwarfs both combatants in Ukraine, suggesting that in any sustained conflict, Chinese drone production could reach truly staggering scales. If Ukraine can manufacture 2.5 million drones annually with a wartime economy and a fraction of China's industrial base, what might China achieve with full mobilization?

Beijing has also observed the limitations of Western support to Ukraine. Despite providing intelligence, weapons, and financial assistance, NATO has been unwilling to directly engage Russian forces or establish a no-fly zone. Chinese planners may calculate that similar hesitation would characterize Western responses to a Taiwan scenario, particularly if China could execute operations quickly, presenting the international community with a fait accompli. The drone war's lessons about the difficulty of force projection and the advantage of defensive operations might be interpreted as favoring the defender of a consolidated position, which China would become once it controlled Taiwan.

China's response to Ukraine's lessons is already visible in military exercises and procurement patterns. The PLA has dramatically increased drone acquisitions, from small tactical systems to large, long-endurance platforms. Chinese defense industries are developing counter-drone technologies, electronic warfare systems, and AI-enabled autonomous capabilities at a pace that alarms Western observers. The integration of drone warfare into PLA doctrine proceeds rapidly, with units training specifically for mass drone operations and counter-drone missions.

Perhaps most significantly, China is developing its own doctrine for drone warfare that may diverge from Ukrainian patterns. With access to advanced artificial intelligence, sophisticated manufacturing, and extensive surveillance infrastructure, Chinese military theorists envision drone swarms operating with greater autonomy, coordination, and sophistication than current systems. The goal is not simply to replicate Ukrainian success but to leap ahead to the next generation of drone warfare, one where human operators become supervisors rather than pilots, where targeting decisions occur at machine speed, and where thousands of drones can execute coordinated operations across vast areas.

The Proliferation Challenge

If the Ukraine conflict were confined to its immediate participants, the strategic implications would be significant but manageable. The reality is far more disturbing: drone warfare technology and tactics are proliferating globally at unprecedented speed. The barriers to entry are remarkably low, the components widely available, and the operational concepts spreading through open-source channels, social media, and international networks of volunteers who participated in Ukraine's defense.

Non-state actors have watched Ukraine's drone war with particular interest. Terrorist organizations, insurgent groups, and militant factions across the Middle East, Africa, and Asia now understand that strategic impact is achievable without air forces, without billion-dollar weapons programs, and without the industrial infrastructure previously required for military power projection. A determined group with modest resources can acquire or build drones capable of striking targets hundreds of kilometers away, of overwhelming local defenses through numbers, and of sustaining campaigns that would have required nation-state resources just a decade ago.

The Syrian civil war and conflicts in Yemen, Libya, and Nagorno-Karabakh provided early demonstrations of military drone effectiveness, but these involved relatively sophisticated, expensive systems like Turkish Bayraktars or Iranian Shahed-series drones requiring state-level procurement and operation. Ukraine has shown that far cheaper, simpler systems can be equally or more effective when produced and deployed at scale. The FPV drones that became workhorses of the Ukraine conflict can be built in garage workshops using readily available components. The technical knowledge required is disseminating rapidly through online communities, instruction videos, and the transfer of expertise as foreign volunteers return home from Ukraine.

Regional conflicts are already being transformed by drone proliferation. The Azerbaijani victory over Armenian forces in Nagorno-Karabakh in 2020 demonstrated the decisive impact of drones in conventional warfare, but that conflict featured relatively expensive Turkish systems. Future regional wars will likely see both sides fielding thousands of cheap, expendable drones, fundamentally altering the calculus of military operations. Nations that historically relied on superior training, better equipment, or defensive terrain to offset numerical disadvantages face new threats from adversaries who can leverage inexpensive drone production to neutralize these advantages.

For counter-terrorism efforts and stability operations, the proliferation of drone technology poses acute challenges. Western militaries have grown accustomed to operating in environments where they maintain air superiority and technological dominance. Small outposts, forward operating bases, and patrol routes were vulnerable to ambush and indirect fire, but the overall strategic picture favored forces with advanced capabilities and robust logistics. When insurgent groups can deploy drones for reconnaissance, targeting, and direct attack, this calculus changes dramatically. The U.S. withdrawal from Afghanistan occurred before widespread insurgent drone use materialized, but future interventions will face adversaries equipped with capabilities that simply did not exist in earlier conflicts.

International efforts to control drone proliferation face formidable obstacles. Unlike nuclear weapons or advanced missile systems, drones can be built from commercial components using publicly available designs. Export controls can restrict the most sophisticated systems, but these represent a small fraction of militarily useful platforms. The dual-use nature of drone technology, with commercial applications from agriculture to infrastructure inspection to package delivery, makes comprehensive restriction impossible without severely impacting legitimate economic activity. Even if major powers agreed on strict export controls, enforcement would be difficult and circumvention relatively straightforward.

The genie is out of the bottle, and no realistic scenario returns it. The strategic implications of drone proliferation will unfold over years and decades, but the trajectory is clear. Future conflicts at every scale, from great power wars to regional disputes to counter-insurgency operations to terrorist attacks, will feature drones prominently. The military organizations, governments, and alliances that adapt most effectively to this reality will possess significant advantages. Those that fail to adapt, that cling to legacy systems and doctrine, or that underestimate the transformative impact of cheap, proliferated drone technology risk devastating surprises when conflict comes.

Rewriting the Rules

The strategic implications of Ukraine's drone war extend beyond military planning into the legal, ethical, and diplomatic frameworks that govern armed conflict. International humanitarian law, developed over centuries and codified in Geneva Conventions and subsequent protocols, rests on assumptions about warfare that drone proliferation challenges fundamentally. The principle of distinction between combatants and civilians becomes harder to enforce when quadcopters can strike anywhere. Proportionality calculations shift when the attacking weapon costs five hundred dollars and the defensive response costs millions. The requirement for human judgment in targeting decisions faces pressure from autonomous systems that can outperform human operators in speed and accuracy.

Ukraine has documented numerous instances of Russian drones deliberately targeting civilians, dropping landmines in populated areas, and striking clearly marked humanitarian facilities. These are war crimes under existing law, but prosecution and accountability become complex when operations are distributed across thousands of operators, many of whom fight remotely from positions far behind front lines. The distance between operator and effect, always present in long-range warfare, becomes even more pronounced with drones. The psychological and moral implications of this distance remain poorly understood and inadequately addressed in military training and legal frameworks.

The international community faces difficult questions about how to regulate drone warfare without undermining legitimate defense capabilities or creating unenforceable rules that only law-abiding nations follow. Some nations push for restrictions on autonomous weapons, arguing that machines should not make life-and-death decisions without human oversight. Others contend that autonomous systems can be more precise and discriminating than human operators under stress, potentially reducing civilian casualties rather than increasing them. The debate is urgent because technology is advancing far more rapidly than international consensus can form.

Strategic stability, for generations maintained through arms control agreements, confidence-building measures, and diplomatic engagement, requires new frameworks for the drone age. When nations can manufacture millions of drones annually, how does verification work?

Traditional arms control focused on delivery systems, nuclear warheads, and major platforms that could be counted and monitored. Drones produced in distributed facilities from commercial components defy easy verification. A nation could hide substantial production capacity or rapidly surge manufacturing in ways that would have been impossible with previous weapons systems.

The Ukraine conflict has rewritten strategic assumptions faster than doctrine can adapt. Military organizations worldwide are racing to absorb lessons while the fighting continues, knowing that their conclusions will shape conflicts for decades to come. The implications ripple outward from tactical adjustments through operational concepts to grand strategy. Deterrence, defense, power projection, alliance structures, arms control, and the fundamental calculus of war and peace all require rethinking in light of what Ukrainian and Russian forces have demonstrated.

The strategic revolution is not complete. Ukraine's drone war continues evolving, with new technologies, tactics, and systems emerging constantly. What remains clear is that the warfare paradigm has shifted fundamentally and irrevocably. Nations, alliances, and militaries that recognize this reality and adapt decisively will shape the strategic environment of the coming decades. Those that fail to adapt will find themselves at a potentially catastrophic disadvantage when future conflicts test their assumptions. The lessons are being written in Ukrainian skies, and the world is watching, learning, and preparing for a future where drones are not merely a feature of warfare but its defining characteristic.

Chapter 26

Ukraine as Drone Superpower

In February 2022, Ukraine possessed virtually no indigenous drone manufacturing capacity. The country's military relied on imported systems, primarily Turkish Bayraktar TB2 drones and a handful of other foreign platforms, supplemented by volunteer-procured commercial quadcopters pressed into improvised military service. The nation's defense industry, while capable in certain areas, had no experience producing unmanned aerial systems at scale. Three and a half years later, Ukraine stands as one of the world's leading drone powers, manufacturing millions of systems annually, pioneering combat-tested technologies, and teaching NATO forces how to defend against and employ the weapons that have come to define modern warfare. This transformation, forged under the existential pressure of invasion and executed at breathtaking speed, represents one of the most remarkable defense industrial achievements of the twenty-first century.

From Importer to Innovator

The early months of the war revealed Ukraine's drone dependency with painful clarity. Turkish Bayraktars achieved spectacular successes against Russian armored columns, becoming symbols of Ukrainian resistance and subjects of a patriotic song that swept the nation. Yet Ukraine owned only a limited number of these systems, could not produce them domestically, and found itself at the mercy of Turkish export decisions and production capacity. When drones were destroyed in combat or degraded through wear, replacement meant diplomatic negotiations, international purchases, and supply chains vulnerable to interdiction and political pressure.

Commercial drones from DJI and other manufacturers filled gaps in reconnaissance and artillery spotting, but these too came from abroad, primarily from China. Volunteer networks mobilized impressively, raising funds through crowdsourcing campaigns and purchasing consumer quadcopters by the thousands from international retailers. The Angry Birds group and similar organizations demonstrated remarkable creativity and dedication, but the fundamental problem remained: Ukraine depended entirely on external sources for the tools that were proving decisive on the battlefield.

China's decision to restrict DJI Mavic exports to Ukraine, responding to political pressure and concerns about military use of commercial systems, crystallized the vulnerability. Suddenly, one of Ukraine's primary reconnaissance platforms became difficult to acquire in the quantities needed. The restriction could have crippled Ukrainian operations at a critical moment, forcing commanders to make difficult decisions about prioritizing scarce resources and accepting gaps in coverage.

Instead, the restriction catalyzed transformation. If Ukraine could not import the drones it needed, it would build them. What followed was an industrial mobilization unlike anything seen in Europe since World War II, compressed into a timeframe that would have seemed impossible to traditional defense procurement experts. The nation's engineers, entrepreneurs, and technicians, many of whom had backgrounds in software development, electronics manufacturing, and small-scale production, pivoted to drone manufacturing with urgency born of survival necessity.

The first domestic systems were crude by conventional standards. Assembled in apartments, garages, and small workshops using whatever components could be sourced, these early Ukrainian drones prioritized function over refinement. They flew, they carried cameras or warheads, they could be controlled by operators who received minimal training. That they worked at all was achievement enough. That they worked well enough to be useful in combat was remarkable. That they could be produced in numbers sufficient to impact operations was transformational.

Innovation emerged not from design bureaus and research laboratories but from the battlefield itself. Operators identified problems, communicated needs, and often implemented solutions themselves. The feedback loop between user and manufacturer, which in peacetime procurement can stretch across years and multiple bureaucratic layers, collapsed to days or even hours. A drone unit discovering that a particular frequency was jammed would report the issue, and manufacturers would adjust designs accordingly. When FPV pilots needed longer range, engineers increased battery capacity and optimized power management. When electronic warfare intensified, fiber-optic guidance solutions emerged within months rather than the years traditional development cycles would require.

This iterative, combat-driven development process produced systems that, while less polished than their foreign counterparts, were often more effective for the specific conditions of Ukraine's war. They were adapted to local electronic warfare environments, designed for the ranges and payloads Ukrainian tactics required, and optimized for rapid production using available supply chains. Most importantly, they could be manufactured in Ukraine, by Ukrainians, at scales that met the enormous consumption rates of high-intensity conflict.

The Manufacturing Miracle

By late 2024, Ukraine had achieved what many analysts considered impossible: complete domestication of FPV drone production. The country that had manufactured zero military drones in early 2022 was now producing 1.5 million FPV systems annually, with capacity expanding rapidly. TAF Drones, the largest manufacturer, operated facilities producing 40,000 units per month, an output valued at over one billion dollars annually. Dozens of smaller manufacturers contributed thousands more, operating from facilities ranging from proper factories to converted apartment buildings to networks of distributed workshops.

The manufacturing ecosystem that emerged bore little resemblance to traditional defense production. Where conventional weapons manufacturing concentrates in large, specialized facilities with extensive security and quality control infrastructure, Ukrainian drone production is distributed across hundreds of locations. This distribution reflected both practical necessity, concentrated facilities would be vulnerable to Russian strikes and the nature of drone manufacturing itself. The systems were relatively simple, assembled from components that could be produced separately and combined in final assembly. A single centralized factory was neither necessary nor desirable.

Quality control in this distributed system posed significant challenges. With production scattered across so many facilities, maintaining consistent standards required new approaches. Ukrainian manufacturers developed modular testing protocols, where components could be validated at various stages rather than relying solely on final inspection. Operators at the front provided brutal but effective quality feedback, reporting systems that failed in combat and creating reputation incentives for manufacturers to maintain standards. The market mechanism, rarely applicable in defense procurement, worked with remarkable efficiency when units could choose suppliers based on performance.

The regulatory environment facilitated rapid scaling in ways that would have been impossible in peacetime or in more bureaucratic systems. Ukrainian authorities, recognizing the existential stakes, operated in what amounted to a regulatory gray zone, prioritizing speed and volume over conventional oversight. Environmental regulations, worker safety requirements, quality certification processes, and the countless other rules that govern manufacturing in developed economies were relaxed or ignored entirely. This enabled breathtaking agility but also created sustainability questions for the post-war period.

Financing the expansion required creativity. Traditional defense procurement involves government contracts, competitive bidding, and payment schedules tied to delivery milestones. Ukrainian drone manufacturing often operated on different principles, with crowdfunding, international donations, and direct unit purchases supplementing government funding. The Ukraine Drone Coalition's €2.75 billion commitment provided crucial stability, as did bilateral agreements like Denmark's €1.5 billion package for 2025 production. Yet even these substantial sums represented only a portion of total production, with significant capacity funded through non-traditional channels.

By 2025, Ukraine's annual production capacity reached 2.5 million drones, with President Zelensky announcing plans to scale to 4 million. These numbers, almost incomprehensible in their magnitude, reflected both the consumption rates of sustained high-intensity conflict and Ukraine's success in building the industrial capacity to meet demand. For context, 4 million drones annually means producing more than 10,000 systems per day, every day, without pause. Achieving such output required not just factories but entire supply chains, logistics networks, training pipelines, and maintenance infrastructure built from virtually nothing in under four years.

Teaching NATO How to Fight

Ukraine's drone warfare expertise, purchased at terrible cost, has become its most valuable military export. NATO forces, despite vastly larger budgets and decades of experience in modern warfare, found themselves coming to Ukraine to learn. The alliance that provided training, weapons, and intelligence support in the war's early stages increasingly recognizes that Ukraine possesses knowledge and capabilities in drone warfare that exceed anything available in NATO arsenals or doctrine.

This role reversal represents a profound shift in military-technological relationships. Historically, Western militaries pioneered new technologies and operational concepts, which then diffused to allied and partner nations through training programs, equipment transfers, and doctrinal guidance.

Ukraine has inverted this flow. NATO officers now rotate through Ukrainian training centers, observing operations, interviewing commanders, and attempting to distill lessons that can be applied to their own forces. The students have become teachers.

What NATO forces are learning from Ukraine goes far beyond technical specifications and operational procedures, though these are important. Ukrainian operators can explain the nuances of flying FPV drones in heavily jammed environments, describe the tactics for defeating fiber-optic guided systems, and demonstrate counter-drone techniques developed through trial and error against Russian swarms. This tactical knowledge is valuable, but the deeper lessons concern mindset, organizational culture, and approaches to innovation that differ fundamentally from NATO norms.

Ukrainian forces embraced radical decentralization, pushing decision-making authority to small unit leaders and even individual operators in ways that NATO's more hierarchical structures struggle to replicate. They accepted failure rates and casualty tolerances that would be politically unacceptable in Western democracies, enabling more aggressive experimentation and faster learning cycles. They operated in regulatory and legal frameworks that prioritized effectiveness over process, moving at speeds that NATO bureaucracies cannot match.

NATO's challenge is determining which Ukrainian lessons are transferable and which are artifacts of Ukraine's specific circumstances. A nation fighting for survival on its own territory, with popular mobilization and existential motivation, can sustain operational tempos and accept risks that professional militaries in limited conflicts cannot. Ukrainian forces improvise because they must; NATO forces prefer methodical planning because they usually can. The question is whether NATO can adopt Ukrainian agility without abandoning the safeguards and oversight that peacetime democracies demand.

The counter-drone warfare expertise Ukraine has developed proves particularly valuable to NATO. Ukrainian air defenses, integrating electronic warfare, kinetic systems, and tactical responses, achieved 93% interception rates against Russian drone swarms by late 2025. These success rates exceeded anything NATO forces had demonstrated in exercises or limited combat operations. Ukrainian operators developed techniques for detecting low-altitude drones, discriminating between threats and decoys, and managing ammunition expenditure when facing saturation attacks.

NATO's eastern flank nations, particularly Poland, the Baltic states, and Romania, engage most intensely with Ukrainian drone warfare lessons. These countries, acutely aware that they would face similar Russian tactics in any future conflict, invest heavily in replicating Ukrainian capabilities and adapting Ukrainian doctrine to their own circumstances. Polish forces conduct regular exchanges with Ukrainian units, Romanian soldiers train on Ukrainian counter-drone systems, and Estonian cyber specialists collaborate with Ukrainian engineers on electronic warfare solutions.

The broader NATO alliance moves more slowly but is moving nonetheless. Major exercises now incorporate scenarios based on Ukrainian experiences, with opposing forces deploying drone swarms modeled on Russian tactics and defending forces attempting to replicate Ukrainian responses. Results are sobering, with NATO units often overwhelmed by the speed, scale, and persistence of simulated drone attacks. These exercises, while humbling, serve their purpose: forcing NATO to confront capability gaps and doctrinal inadequacies before facing these challenges in actual combat.

Defense Industry Transformation

Ukraine's defense industrial base in early 2022 consisted primarily of Soviet-legacy facilities producing artillery shells, armored vehicle components, and maintenance services for aging equipment. The sector employed thousands but focused on sustaining existing systems rather than developing new ones. Innovation occurred at a glacial pace when it occurred at all, constrained by limited budgets, bureaucratic inertia, and the brain drain that had seen many of Ukraine's best engineers emigrate to opportunities abroad.

The war created the necessity, urgency, and resources for comprehensive transformation. Defense industry employment exploded as new companies formed and existing firms pivoted to military production. Engineers who had left for better opportunities in Kyiv's tech sector or abroad returned, motivated by patriotism, professional challenge, or both. Investment flowed from government, international partners, and private sources. Most dramatically, the entire industry operated under wartime conditions that eliminated peacetime constraints and enabled risk-taking that would have been unthinkable before February 2022.

Drone manufacturing emerged as the flagship of this transformation, but the changes extended far beyond unmanned systems. Ukrainian companies developed electronic warfare equipment, created communications systems hardened against jamming, and pioneered counter-drone technologies. The innovation was not limited to hardware; Ukrainian software engineers created command and control systems, developed AI algorithms for target recognition, and built logistics platforms for managing the enormous flow of equipment to front-line units.

The transformation's speed and scope reflected Ukraine's pre-war strengths in software development and electronics manufacturing. Kyiv had been a significant tech hub before the invasion, with thousands of software engineers working for international clients and a thriving startup ecosystem. These skills proved remarkably transferable to defense applications. Engineers accustomed to developing mobile apps or enterprise software found that drone control systems, while more demanding, used similar programming paradigms. Electronics manufacturers who had produced consumer devices discovered that military systems required many of the same capabilities adapted to harsher conditions.

This convergence of commercial technology and military application, which defense analysts call "dual-use technology," became a defining characteristic of Ukrainian defense innovation. Rather than maintaining strict separation between civilian and military sectors, Ukraine effectively mobilized its entire technology industry for the war effort. Companies maintained commercial operations while also producing military equipment, individuals split time between civilian projects and defense work, and knowledge flowed freely across boundaries that in other countries would be much firmer.

The organizational culture that emerged in Ukraine's defense sector diverges sharply from traditional military-industrial norms. Ukrainian drone manufacturers, many led by young entrepreneurs with backgrounds in tech startups rather than defense contracting, brought Silicon Valley ethos to weapons production. They emphasized rapid iteration, user feedback, and continuous improvement over formal requirements and lengthy development cycles. They operated with flat hierarchies where ideas came from anyone regardless of seniority. They accepted that some designs would fail and moved on quickly rather than studying failures endlessly.

This culture clashed at times with military customers accustomed to more formal relationships and with international partners expecting traditional defense industry practices. Yet the results spoke for themselves. Ukrainian manufacturers produced systems faster, adapted more quickly to changing requirements, and achieved better operational performance than would have been possible through conventional approaches. The methods were often chaotic and occasionally produced failures, but the overall trajectory was undeniably upward.

Post-War Export Potential

Even as Ukraine fights for survival, forward-thinking analysts recognize that the nation is building a defense industrial base with extraordinary post-war potential. The knowledge, capabilities, and systems developed under fire will have immense value in global markets increasingly focused on drone warfare and counter-drone defense. Ukraine is positioned to become a major arms exporter, potentially rivaling established defense powers in specific technology areas.

The export potential rests on several foundations. First, Ukrainian systems are combat-proven in ways that virtually no other platforms can claim. Every major Ukrainian drone type has been tested under the most demanding conditions imaginable: against a peer adversary with sophisticated electronic warfare, extensive air defenses, and its own capable drone forces. Nations considering drone acquisitions can review not manufacturer claims or peacetime demonstrations but actual combat footage and operational data from sustained high-intensity conflict. This pedigree is invaluable.

Second, Ukrainian systems are optimized for the kind of warfare that many nations increasingly expect to fight. They operate effectively in denied environments, function despite jamming and GPS spoofing, and suit the constrained budgets of nations that cannot afford American or Israeli defense systems. A country facing regional threats or insurgency challenges may find that Ukrainian drones, designed for brutal attritional warfare, meet their needs better than more sophisticated and expensive Western alternatives.

Third, Ukraine can offer not just hardware but comprehensive doctrine, training, and support based on unmatched operational experience. Purchasing Ukrainian drones comes with access to advisors who have actually employed these systems in combat, training programs developed under fire, and maintenance procedures validated through intensive use. This knowledge ecosystem may prove as valuable as the hardware itself for nations building drone warfare capabilities.

The potential customer base is substantial. NATO members, particularly those on the alliance's eastern flank, are obvious candidates and are already procuring Ukrainian systems in small quantities for evaluation. Middle Eastern nations, many of which have used Turkish and Iranian drones in regional conflicts, represent another major market. Asian countries facing their own security challenges, African nations dealing with insurgencies, and Latin American countries combating organized crime all have potential use cases for Ukrainian drone technology.

Legal and political obstacles will complicate Ukrainian export ambitions. Many nations maintain strict export control regimes governing unmanned systems, concerned about proliferation to non-state actors and the use of drones in human rights abuses. Ukraine will need to develop export control procedures, comply with international agreements, and navigate the political sensitivities of arms sales. The European Union and United States, both crucial Ukrainian supporters, will exert influence over who can purchase Ukrainian systems and under what circumstances.

Economic considerations will also shape export potential. Ukraine will need to balance domestic military requirements against export opportunities, ensuring that production for foreign customers does not shortchange Ukrainian forces. The country will face competition from established defense exporters, who may respond to Ukrainian market entry with aggressive pricing or political pressure. Building international sales networks, providing after-sales support, and establishing credibility as a reliable supplier will require substantial investment and time.

Despite these challenges, the trajectory is clear. Ukraine enters the post-war period with defense manufacturing capabilities far exceeding its pre-war status and competitive with mid-tier defense exporters. The nation that in 2022 could not produce a single military drone will likely be exporting hundreds of thousands of systems annually within a decade, generating substantial revenue and wielding considerable influence through defense relationships.

The Knowledge Base Built Under Fire

Perhaps Ukraine's most valuable asset, more enduring than any specific system or production facility, is the enormous body of knowledge accumulated through three and a half years of intensive drone warfare. This knowledge exists in multiple forms: formal doctrine and training materials, technical specifications and design databases, but most importantly in the expertise of thousands of individuals who learned to operate, manufacture, and counter drones under combat conditions.

Ukrainian drone operators, by late 2025, represent the world's largest and most experienced community of military drone pilots. Thousands of individuals have accumulated hundreds of hours flying FPV systems in contested environments, developing intuitive understanding of how drones behave in wind and weather, how to exploit terrain masking and concealment, and how to evade counter-drone systems. This expertise cannot be replicated through peacetime training or simulation. It reflects real experience making life-and-death decisions, learning from failures, and refining techniques through countless iterations.

The manufacturing knowledge base is equally valuable. Ukrainian engineers and technicians have solved countless problems in materials, propulsion, guidance systems, and integration that would have taken years to address through formal research and development. They understand which components fail in cold weather, how to waterproof systems for mud and rain, which batteries provide the best combination of endurance and cost, and how to assemble drones quickly using minimal tools. This is practical, applied knowledge that cannot be found in textbooks or research papers.

Counter-drone expertise represents another critical knowledge domain. Ukrainian air defense operators, electronic warfare specialists, and tactical commanders have developed a comprehensive understanding of how to detect, track, and defeat drone swarms. They know the signatures that different drone types produce, the frequencies that various guidance systems use, the tactics that attackers employ, and the vulnerabilities that can be exploited. This knowledge, refined through constant engagement with evolving threats, provides Ukraine with capabilities that exceed anything available from defense contractors or research institutions.

The institutional challenge is capturing, preserving, and disseminating this knowledge before it is lost. Individual expertise, while valuable, becomes truly powerful when codified into doctrine, embedded in training programs, and transferred to new generations of operators and engineers. Ukraine has begun this process, establishing training centers, documenting procedures, and creating educational programs, but much work remains. The risk is that critical knowledge resides in individuals who may leave the military, move to the private sector, or simply forget details as time passes.

Ukraine is also grappling with classification challenges. Much of the most valuable knowledge involves tactics, techniques, and procedures that would benefit adversaries if disclosed. Yet overly restrictive classification can prevent sharing with allies, limit commercial applications, and slow the diffusion of knowledge to Ukrainian forces themselves. Finding the right balance between security and dissemination requires judgment that evolves as the conflict continues and as Ukraine's role shifts from embattled defender to post-war defense innovator.

Drone Doctrine Written in Combat

Military doctrine, the formal statements of how to conduct operations, organize forces, and employ capabilities traditionally develops through peacetime analysis, exercises, and study of historical conflicts. The process is deliberate, involving field manuals, staff colleges, and bureaucratic review that can stretch across years. Ukraine has compressed this timeline dramatically, writing drone warfare doctrine in real time as operations continue.

Ukrainian doctrine for drone employment has progressed through distinct phases. Early improvisation gave way to systematized tactics as operators identified what worked and what did not. Organizational structures evolved, with dedicated drone units forming alongside traditional branches. Training standardized as lessons learned spread through the force. By 2025, Ukraine possesses comprehensive doctrine covering drone reconnaissance, FPV attacks, counter-drone operations, electronic warfare, and the integration of drones with conventional forces.

This doctrine differs in important ways from NATO's emerging thinking on unmanned systems. Ukrainian approaches emphasize mass over sophistication, accept higher loss rates than Western forces would tolerate, and assume heavily contested electromagnetic environments as baseline conditions. Ukrainian doctrine is less concerned with protecting individual platforms and more focused on achieving operational effects through volume and persistence. These differences reflect different military cultures, different strategic circumstances, and different resource constraints.

The doctrinal knowledge Ukraine has developed holds value beyond its immediate military application. Academic researchers, defense analysts, and military historians will study Ukrainian drone warfare for decades, mining insights applicable to future conflicts and technological evolution. The conflict provides real-world data on questions that previously could only be addressed through speculation: How do large-scale drone swarms perform in combat? What attrition rates are sustainable? How do human operators and AI systems compare in various tasks? What organizational structures best support drone warfare?

Ukraine's willingness to share doctrinal insights with allies enhances its strategic relationships and amplifies its influence. By hosting foreign observers, participating in international exercises, and publishing (appropriately redacted) lessons learned, Ukraine shapes how other nations approach drone warfare. This doctrinal influence extends Ukrainian power beyond its borders and ensures that the knowledge purchased with Ukrainian blood benefits the broader community of nations defending against aggression.

From Survival to Leadership

The transformation from drone importer to drone superpower occurred not through choice but through necessity. Ukraine did not set out to become a world leader in unmanned systems; it simply had no alternative to rapid innovation and massive scaling. Yet the achievement, regardless of its origins, is undeniable. The nation that in early 2022 depended entirely on foreign drones now manufactures millions annually, pioneers technologies that defense powers study intensively, and teaches NATO forces how to fight in the drone age.

This position brings responsibilities and opportunities. Ukraine must balance domestic defense needs against export potential, share knowledge with allies while protecting sensitive information, and build sustainable defense industries for a post-war future. The country must navigate relationships with traditional defense powers who may view Ukrainian competition with suspicion, manage expectations from partners who expect access to Ukrainian capabilities, and maintain technological edge as adversaries study and copy Ukrainian innovations.

The psychological and cultural shift required to embrace this new role should not be underestimated. Ukraine's self-perception before the invasion was as a country perpetually behind, catching up to Western standards, dependent on more advanced nations for technology and security. The drone warfare transformation has forced a fundamental reassessment. Ukraine now possesses capabilities that more established powers lack, expertise that cannot be purchased or learned except through the crucible of high-intensity combat, and industrial capacity built from almost nothing in record time.

This shift from dependence to leadership will shape Ukraine's post-war trajectory in ways that extend far beyond defense manufacturing. A nation that can innovate under existential pressure, mobilize entire industries in months, and out-compete established powers in specific technology domains has demonstrated capacity for achievement in other areas. The confidence, organizational capability, and human capital developed through the drone warfare transformation will enable Ukraine's broader post-war reconstruction and integration into European structures.

The drone superpower status, earned through suffering and sacrifice, represents more than military achievement. It symbolizes Ukraine's resilience, creativity, and determination to not merely survive but to transform adversity into strength. The knowledge base built under fire will endure long after the last drone flies, shaping how the world understands and conducts warfare in the decades ahead. Ukraine's contribution to this transformation, purchased at terrible cost, ensures that the nation will never again face threats without the means to defend itself and the knowledge to teach others how to resist aggression.

PART VI: THE HUMAN DIMENSION

Chapter 27

The Drone Operators

The transformation of warfare through unmanned systems is fundamentally a human story. Behind every drone strike, every reconnaissance mission, every successful interception lies an individual operator making split-second decisions with lethal consequences. These operators, many of whom were civilians mere months before donning uniforms, represent a new category of warrior whose experiences differ profoundly from those of traditional combatants. They kill from kilometers away, watching death unfold through electronic displays. They operate from relative safety yet bear psychological burdens that military medicine is only beginning to understand. They are gamers turned soldiers, technicians turned killers, and their stories illuminate both the promise and peril of warfare's evolution.

Dmytro's Journey: From Software Developer to FPV Pilot

Dmytro's life before February 2022 revolved around code, not combat. At twenty-eight, he worked for a Kyiv software firm developing mobile applications, earning a comfortable salary that afforded a pleasant apartment in the city's Podil district, regular vacations to Turkey and Poland, and aspirations of eventually moving to Berlin or Amsterdam for better opportunities. He had completed mandatory military service years earlier, an experience he described as tedious and largely pointless, and harbored no particular martial ambitions or interests.

The invasion shattered his comfortable trajectory. Like millions of Ukrainians, Dmytro watched the first dawn of Russian bombardment with disbelief that quickly hardened into rage and determination. Within days, he had volunteered for territorial defense forces, expecting to receive an assault rifle and basic training before being sent to dig trenches or man checkpoints. Instead, a recruiter noticed on his intake form that Dmytro flew racing drones as a hobby, a passion he had pursued for three years, competing in local events and posting flight videos online.

"They asked if I could fly with FPV goggles," Dmytro recalls, sitting in a dugout position twenty kilometers from the front line in Ukraine's east. "I said of course, I had been racing for years. They told me racing was over, now I would hunt." Within two weeks of volunteering, Dmytro found himself in a hastily organized training program learning to fly military drones. The transition from racing through obstacle courses to targeting Russian vehicles required significant adjustment, but the fundamental skills of spatial awareness, hand-eye coordination, managing latency and flying through interference translated directly.

His first kill came on a gray April morning in 2022. The target was a Russian logistics truck parked near a tree line, identified by reconnaissance operators and confirmed by artillery spotters. Dmytro launched his FPV drone, a commercial quadcopter modified to carry a small explosive charge, and flew toward the coordinates. "I was shaking so badly I nearly crashed twice before reaching the target," he remembers. "When I hit the truck and saw the explosion through my goggles, I felt sick. I had to step away and vomit. My hands wouldn't stop shaking for hours."

By late 2025, Dmytro had logged over three thousand flight hours and eliminated dozens of targets: vehicles, artillery pieces, electronic warfare stations, command posts, and enemy personnel. The shaking stopped after the first few missions. The nausea faded. What replaced these visceral reactions is something harder to define, a professional detachment, a mechanical process of target identification, approach planning, and execution. "I don't think about the people inside anymore," he says quietly. "If I did, I couldn't do the job. I focus on the technical aspects: angle of approach, wind compensation, timing the detonation. It's like solving a coding problem, except the solution is an explosion."

The psychological compartmentalization that allows Dmytro to function comes at cost. He sleeps poorly, startled awake by sounds that remind him of drone motors. Alcohol consumption has increased steadily, a pattern he recognizes intellectually but struggles to control. Relationships with friends and family have become strained; he finds civilian concerns trivial and cannot relate to their everyday worries about work deadlines or apartment repairs. "They live in a different world," he explains. "I've killed more people than most soldiers in history got close to. How do I explain that to someone worried about their promotion?"

Oksana's Story: Artillery Spotter Turned Hunter

Oksana joined the military in 2021, before the invasion, serving as a communications specialist with an artillery brigade. The work was technical and safe, involving radio equipment maintenance and signal management far behind any potential front line. When war came, her unit found itself engaged immediately, supporting desperate defensive actions against Russian advances toward Kyiv.

The introduction of reconnaissance drones transformed Oksana's role. Her unit received several commercial DJI Mavics in March 2022, donated by volunteers and delivered without instructions or training. Officers assumed that young soldiers, being "digital natives," would figure out how to operate them. Oksana, who had never flown a drone in her life, spent an evening reading online tutorials and watching YouTube videos before attempting her first flight the next morning.

"I crashed it immediately," she laughs, recounting the experience. "Just flew it straight into a tree within thirty seconds. Everyone was furious because these drones cost thousands of dollars and we only had three. But I kept trying, and by afternoon I could keep it airborne and moving in the right direction. Within a week, I was doing reconnaissance missions and calling in artillery strikes."

The transition from communication specialist to combat drone operator happened without formal orders or official recognition. Oksana simply possessed a valuable skill and willingness to employ it, which in the chaotic early months of the war was sufficient. She flew dozens of missions, providing targeting data that enabled her artillery unit to destroy Russian vehicles, ammunition depots, and troop concentrations. The work felt abstract, distant; she identified targets and transmitted coordinates, but rarely witnessed the actual strikes or their effects.

This abstraction ended when Oksana transitioned to FPV operations in late 2023. Her reconnaissance experience and demonstrated competence with unmanned systems made her an obvious candidate when her brigade began forming dedicated drone units. FPV operations, unlike reconnaissance, involved direct engagement. Oksana would pilot the weapon to target herself, watching the final seconds through her goggles as the drone impacted.

The first-person perspective changed everything. "With artillery spotting, I was giving information to other people who made the decision to fire," Oksana explains. "With FPV, I am the weapon. I see the target, I decide to engage, I guide the drone into impact. There's no distance, no intermediary. It's intimate in a way that feels wrong."

The intimacy has proven particularly difficult when targeting personnel. Vehicles and equipment present clear military objectives, but individual soldiers, even enemy soldiers, register differently in Oksana's consciousness. She describes watching through her goggles as soldiers attempt to evade incoming drones, running, diving, sometimes firing uselessly at the small, fast-moving targets. "They're trying to live," she says simply. "You see their desperation, their humanity. Then you kill them anyway because that's the mission."

Oksana sought psychological counseling after several particularly difficult missions in early 2025. The military's mental health resources, while improved from the war's early days, remain inadequate for the scale of need. She met with a counselor twice before the therapist rotated to another unit. The sessions helped somewhat, providing space to articulate feelings she could not share with fellow soldiers or family, but offered no resolution to the fundamental moral tensions of her work.

"People ask if I have PTSD," Oksana reflects. "I don't think that's the right framework. I'm not traumatized by what I've experienced; I'm troubled by what I've done. There's a difference. I chose this. Every target I've engaged, I could have refused. But I didn't refuse because the mission required it, because we're defending our country, because if I don't kill them they'll kill us. Understanding why I do it doesn't make it easier to live with."

Training the Next Generation

Major Volodymyr oversees one of Ukraine's primary drone operator training programs, a facility established in 2023 that has produced over four thousand qualified FPV pilots. A career officer with twenty years of military experience, he came to drone warfare as a skeptic, viewing unmanned systems as supporting assets rather than primary combat platforms. Three years of war thoroughly revised his assessment.

"In the beginning, we treated drone operators as technicians," Major Volodymyr explains, walking through the training center's classrooms and flight simulator stations. "We gave them a few days of basic instruction and sent them forward. The attrition rate was terrible, both in people and equipment. We lost operators to enemy fire because they didn't understand tactical considerations. We lost drones because pilots lacked fundamental skills. We had to professionalize the entire enterprise."

The current training program runs eight weeks for FPV operators, significantly longer than the hasty instruction of 2022 but still compressed compared to traditional military pilot training. Candidates arrive with varied backgrounds; some are gamers with extensive experience in simulators and racing drones, others are soldiers transferring from other specialties, and still others are recent recruits with no relevant experience at all.

The first two weeks focus on fundamental flight skills using simulators. Students spend hours each day practicing basic maneuvers of takeoffs, landings, navigation and altitude control in virtual environments that replicate the physics of actual drones without risking expensive hardware. The simulators, many donated by international supporters or built by Ukrainian software developers, provide immediate feedback and allow rapid iteration. Students can crash hundreds of times without consequence, learning from mistakes that would be costly or impossible to repeat with physical systems.

Weeks three and four introduce actual drones, beginning with older or damaged units that can be sacrificed to learning. Students fly increasingly complex missions in controlled areas, developing muscle memory and situational awareness that simulators cannot fully replicate. Instructors observe closely, identifying natural talent and correcting dangerous habits before they become ingrained.

The curriculum's second half addresses combat operations. Students learn tactical considerations: how to approach targets to minimize detection, how to compensate for wind and weather, how to navigate in GPS-denied environments, and how to evade counter-drone fire. They study electronic warfare, understanding which systems they'll face and how to recognize jamming. They receive instruction in target identification, learning to distinguish military vehicles from civilian equipment and to recognize high-value targets.

The psychological preparation proves most challenging. Major Volodymyr and his instructors recognize that they're training students to kill, often at very close range through first-person perspective, and that this experience will affect them profoundly. The program includes discussions of combat stress, moral injury, and coping mechanisms, but these sessions feel inadequate to the task. "How do you prepare a twenty-year-old to watch someone die through their goggles?" Volodymyr asks rhetorically. "We do our best, but ultimately they learn through experience, and that learning is painful."

The program graduates students with basic proficiency, but real expertise develops only through combat operations. The training center maintains contact with graduates, collecting feedback on what skills proved most valuable and what gaps in preparation became apparent under fire. This feedback loop continuously refines the curriculum, but the fundamental challenge remains: compressing years of skill development into weeks and preparing people psychologically for experiences that no training can fully simulate.

The Moral Weight of Remote Killing

The psychological impact of drone warfare diverges from traditional combat in ways that military psychology is only beginning to understand. Conventional wisdom holds that distance from combat reduces psychological trauma, artillerymen firing at targets beyond visual range experience less PTSD than infantry engaged in close combat. Yet drone operators, despite operating from relative safety, often report significant psychological distress.

The paradox reflects drone warfare's unique characteristics. Operators watch their targets intimately through high-resolution cameras, seeing details that soldiers in traditional combat rarely observe. They witness the moments before death with clarity that would be impossible from a distance or through the chaos of ground combat. Yet they experience this intimacy without physical risk, without the adrenaline and fear that characterize direct engagement, and often from positions far removed from any combat zone.

"I've killed people while sitting in a comfortable chair, drinking coffee, in a building with heat and electricity," says Maksym, an FPV operator with over two years of combat experience. "Sometimes I finish a mission, confirm the kill, and then walk to the mess hall for lunch. The dissonance is overwhelming. I just ended someone's life, and twenty minutes later I'm eating borscht and joking with friends. It doesn't feel real, but it's completely real. I don't know how to process that."

The moral weight intensifies when operators engage personnel rather than equipment. Destroying a tank or artillery piece feels like degrading enemy combat capability, a clear military objective with minimal moral ambiguity. Targeting individual soldiers, especially those who are retreating, wounded, or otherwise non-threatening in the moment, raises difficult questions that operators must resolve individually.

Ukrainian military regulations provide rules of engagement, but these guidelines cannot address every situation operators encounter. What about a soldier who has abandoned his weapon and is clearly attempting to surrender? What about wounded personnel awaiting evacuation? What about groups that include individuals who might be medics or chaplains? Operators make these determinations in seconds, with imperfect information, under pressure to achieve mission objectives.

Some operators develop strict personal rules to manage moral complexity. They engage only military equipment or soldiers actively engaged in combat operations. They avoid targeting individuals who are clearly non-combatants or hors de combat. Others take a more expansive view, arguing that any Russian soldier in Ukraine is a legitimate target regardless of their immediate actions. These philosophical differences reflect broader debates about proportionality, necessity, and the ethics of warfare that have no simple resolution.

The documentation inherent in drone warfare adds another psychological dimension. Most FPV strikes are recorded, creating permanent visual records of kills. These videos serve operational purposes, confirming targets destroyed, analyzing tactics and training future operators but they also create lasting artifacts of death that operators cannot escape. Unlike soldiers whose combat memories may fade or distort over time, drone operators can review their kills in high definition, watching repeatedly if they choose or if required for debriefing.

Some operators report obsessive reviewing of footage, seeking to understand or justify their actions, or tormented by details they missed during actual operations. Others refuse to watch their recorded missions after initial debriefing, wanting to minimize their exposure to images they cannot forget. The choice of how to engage with this documentation becomes part of each operator's psychological management strategy.

Effectiveness Versus Moral Injury

The tension between combat effectiveness and psychological wellbeing creates difficult decisions for commanders and military planners. Drone operators, particularly highly skilled pilots with dozens or hundreds of successful engagements, represent valuable assets whose expertise takes months to develop and cannot be easily replaced. Yet these same individuals, precisely because of their extensive combat experience, often bear the heaviest psychological burdens.

Ukrainian forces have implemented rotation policies intended to prevent burnout, pulling operators off frontline duties periodically for rest and recovery. The effectiveness of these rotations varies. Some operators report that time away from operations provides crucial psychological relief, allowing them to reconnect with normal life and restore mental equilibrium. Others find that removal from their units and missions increases anxiety, leaving them feeling useless while comrades continue fighting. The optimal rotation schedule appears highly individual, varying based on personality, coping mechanisms, and external support networks.

The military has expanded psychological support services throughout the war, recognizing that mental health casualties can be as debilitating as physical injuries. Combat stress reaction teams now operate near front lines, providing immediate intervention for personnel showing acute symptoms. Longer-term counseling is available at rear-area facilities. Peer support groups, where operators can discuss experiences with others who share similar burdens, have emerged organically and receive official encouragement.

Yet resources remain inadequate for the scale of need. Ukraine's mental health infrastructure was limited before the war and has been further strained by three years of high-intensity conflict affecting millions of people. Military psychologists are in particularly short supply, and those available often lack specialized training in the unique challenges of drone warfare. Most support comes from general combat stress counseling adapted to drone operators' specific circumstances, a helpful approach but not fully adequate to the distinct nature of remote killing.

The long-term psychological impact of extensive drone operations remains unknown. Unlike traditional combat, which military medicine has studied for generations, large-scale use of FPV drones in high-intensity conflict is unprecedented. Current understanding comes from limited research on U.S. Predator and Reaper operators, who typically conducted strikes against asymmetric targets from bases in the United States, a very different context from Ukrainian operators engaged in peer conflict from forward positions.

Some Ukrainian drone operators report positive psychological outcomes from their service despite the killing they've done. They express pride in contributing to national defense, satisfaction at proven competence in demanding work, and strong bonds with unit members. These positive factors may provide psychological resilience that offsets some moral injury, though longitudinal studies will be needed to understand long-term trajectories.

The New Face of Warfare

Walking through Ukrainian drone units, one encounters a remarkably diverse group. Young men who worked in IT development sit beside middle-aged women who ran small businesses. Former professional gamers operate alongside farmers and teachers. Educational backgrounds range from advanced technical degrees to limited formal schooling. What unites them is aptitude for drone operations and willingness to serve, not traditional military demographics.

This diversity reflects drone warfare's low barriers to entry relative to other combat specialties. Flying FPV drones requires hand-eye coordination, spatial awareness, and comfort with technology, skills distributed widely across populations and not correlated with physical strength, age, or gender. Unlike infantry operations, which demand physical fitness and endurance, or armor operations, which require strength to manage heavy equipment, drone piloting can be performed by anyone with the necessary cognitive abilities and training.

The democratization of lethal capability challenges traditional military culture in subtle ways. Elite combat units historically prided themselves on exclusivity, on physical and mental standards that few could meet. Drone operations create new elites based on different criteria such as reaction time, technical proficiency and hours in simulators that don't align with conventional martial virtues. This shift generates tension between traditional military branches and emerging drone specialists, with questions about prestige, command authority, and career progression still being negotiated.

Some drone operators embrace their role as representing warfare's future, seeing themselves as pioneers of new combat paradigms. Others feel ambivalent about their position, uncomfortable with the distance between themselves and traditional combat while recognizing that their contribution is essential. "Real soldiers fight with rifles," one operator says, expressing a common sentiment. "I sit in a dugout flying toy airplanes with grenades attached. It's effective, but it doesn't feel like soldiering."

This self-perception issue affects recruitment and retention. Some highly skilled operators leave military service for better-paying positions in defense industry or civilian technology sectors, viewing their military time as necessary service rather than career path. Others commit to long-term military careers, recognizing that drone warfare represents the future and that their expertise will only become more valuable. Ukrainian military leadership struggles to create career pathways and incentive structures that retain skilled operators while maintaining overall force structure.

Living with What They've Done

Years after the war ends, Ukraine will contain thousands of individuals who spent months or years killing remotely, watching death unfold through electronic displays, and bearing psychological burdens that civilian society may struggle to understand or accommodate. These veterans will need support, recognition, and integration into peacetime society in ways that Ukraine's existing veteran services are ill-prepared to provide.

Some operators have begun preparing for this transition already, even as fighting continues. They attend educational programs when off duty, pursue online degrees, and make plans for post-war careers. Others focus entirely on the present, unable or unwilling to contemplate life beyond the conflict. The psychological literature on combat veterans suggests that having purpose and direction after military service significantly improves outcomes, but creating that purpose while still engaged in high-intensity operations presents obvious challenges.

Civilian understanding of drone warfare remains limited and often misinformed. Popular culture portrays drone operators as detached gamers, warfare as video games, killing as bloodless and morally uncomplicated. Operators know differently. They understand that the screens show real death, that their decisions have permanent consequences, and that distance from physical danger does not eliminate moral weight. When they eventually return to civilian life, many worry about explaining their experiences to people who hold these oversimplified views.

"People will ask what I did in the war," says Andriy, an FPV pilot with two years of combat experience. "What do I tell them? That I killed dozens of people from a comfortable position? That I watched men die through video goggles? That I'm good at it, that I take pride in my skill? They won't understand. They'll either think I'm a hero or a monster, and neither is right. I'm just someone who did what needed doing, and who will live with it forever."

The question of how Ukraine will memorialize and honor its drone operators remains open. Traditional military honors celebrate valor, sacrifice, and physical courage, concepts that fit awkwardly with remote operations conducted from relative safety. Yet the contribution of drone operators to Ukraine's defense is undeniable and deserving of recognition. Finding forms of commemoration that honor their service while acknowledging the distinct nature of their experience will challenge Ukrainian society.

What is clear is that the thousands of Ukrainians who became drone operators during this conflict represent both the evolution of warfare and enduring questions about humanity's relationship with technology and violence. They are warriors for a new age, effective and troubled, skilled and scarred. Their stories, uncomfortable and complex, illuminate both the promise of technological advantage and the costs that remain stubbornly human. Long after the drones stop flying, their experiences will resonate through Ukrainian society and inform how future generations understand the war that transformed their nation and the world.

Chapter 28

Life Under the Drones

For millions of Ukrainians, the war is not measured in territory lost or gained, in diplomatic negotiations or military casualties reported on the evening news. It is measured in sleepless nights interrupted by air raid sirens, in the distinctive buzz of Shahed drones overhead, in the calculations about whether to descend to the shelter or risk staying in bed. It is measured in the psychological weight of constant vigilance, in children who have known nothing but war, and in the fundamental rewiring of daily life around the reality of attack from above. This is the civilian experience of drone warfare, less dramatic than battlefield accounts but no less consequential in shaping how a nation endures, adapts, and ultimately survives.

The Nightly Ritual

In Kyiv, Lviv, Odessa, Dnipro, and cities across Ukraine, the pattern has become grimly routine. As darkness falls, anxiety rises. Families finish dinner earlier than they once did, mindful that meal preparation might be interrupted by sirens. Evening entertainment shifts toward activities that can be abandoned quickly, streaming video rather than theater performances, reading at home rather than dining out. Bags are packed near doors: documents, medications, water bottles, flashlights, phone chargers. Children sleep in clothes rather than pajamas, ready for rapid descent to shelters.

The sirens, when they come, trigger responses that have evolved from panic into practiced ritual. In the war's early months, every alert prompted immediate shelter-seeking, with families crowding into basements, metro stations, and purpose-built facilities. By 2025, the calculation has become more nuanced and more troubling. Ukrainians have learned to gauge risk based on the alert's characteristics, on social media reports from eastern regions where drones are first detected, and on intuition built from hundreds of previous alerts.

"You learn to read the situation," explains Iryna, a Kyiv teacher and mother of two. "If the alert sounds and my phone shows drones still far to the east, over Poltava or Cherkasy, I know we have time. If explosions are already audible, if the trajectory suggests they're heading our way, we go down immediately. It's a calculation you never thought you'd have to make, balancing the threat of death against the disruption of constantly sheltering."

The shelters themselves have evolved from improvised spaces into semi-permanent communities. Kyiv's metro stations, serving as shelters throughout the war, have taken on characteristics of underground villages during extended alerts. Regular occupants claim preferred spots, bringing folding chairs, blankets, and thermoses. Elderly residents who struggle with stairs sometimes descend during early evening and remain until morning, preferring hours of discomfort to repeated climbs. Children do homework by flashlight or phone screen, their ability to study amid chaos a testament to human adaptability and a tragedy of normalized trauma.

Not everyone shelters. The decision to ignore alerts, to remain in apartments despite sirens and interception fire overhead, becomes more common as the war extends. Alert fatigue, the psychological exhaustion from constant vigilance leads some residents to accept risk rather than disrupt lives further. "I can't spend every night in the basement," says Oleksandr, a Kyiv software engineer. "I have work deadlines, I need to sleep in a real bed, I need to maintain some semblance of normal life. Maybe it's foolish, maybe I'm risking my life, but this is the choice I've made."

This choice becomes more dangerous when drones penetrate deep into cities. The introduction of fiber-optic guided FPV systems in 2024 extended Russian strike capabilities beyond infrastructure targets to essentially anywhere in Ukrainian territory. Drones have struck apartment buildings, markets, and public squares far from obvious military objectives. The randomness of some attacks, whether intentional terror or guidance failures, means that nowhere feels truly safe, that sheltering provides only marginal protection, and that luck becomes a factor in survival.

Economic Impact and Daily Disruption

The economic consequences of constant drone attacks extend far beyond the direct destruction of infrastructure. Every alert disrupts commerce, halts transportation, and reduces productivity. Factories shut down production lines when sirens sound. Hospitals pause non-emergency procedures. Schools dismiss students or move classes to basements. The cumulative effect represents billions in lost economic activity, a hidden cost of drone warfare that compounds the visible damage to power plants and fuel depots.

For businesses, the disruption creates impossible choices. Maintaining operations during alerts risks employee lives; ceasing operations whenever sirens sound risks bankruptcy. Different enterprises resolve this dilemma differently. Essential services such as hospitals, utilities and emergency response maintain operations with personnel sheltering in place when possible. Retail establishments close immediately, prioritizing employee and customer safety over sales. Manufacturing facilities make risk calculations based on their specific circumstances, sometimes continuing production in sectors where shutdown and restart impose substantial costs.

The insurance industry has essentially collapsed for many risk categories. Standard policies exclude war damage, and special war risk insurance has become prohibitively expensive or entirely unavailable. Property owners bear unreimbursed losses, creating cascading economic effects. A destroyed apartment building represents not just housing loss but complete financial ruin for owners who cannot rebuild, cannot sell damaged property, and often still carry mortgage obligations on uninhabitable units.

The energy sector provides the clearest example of strategic economic warfare through drones. Russia's sustained campaign against Ukrainian power generation and distribution infrastructure, conducted primarily with Shahed drones supplemented by missiles, has systematically degraded electricity supply. Winter 2024-2025 brought rolling blackouts, scheduled outages, and emergency restrictions as damaged facilities struggled to meet demand. Ukrainian households learned to structure activities around electricity availability: charging devices during powered hours, cooking meals when possible, rationing heating to preserve generation capacity.

These outages ripple through every aspect of daily life. Refrigeration becomes unreliable, changing food storage and consumption patterns. Internet connectivity, essential for remote work and maintaining connections with displaced family members, becomes intermittent. Medical equipment requiring electricity forces difficult decisions about treatment priorities. The elderly and disabled, particularly those depending on elevators in high-rise buildings, face isolation when power failures trap them in apartments.

Adaptation takes many forms. Solar panels and battery systems, once luxury items, become survival equipment. Sales of generators surge despite high costs and fuel requirements. Ukrainians share electricity, running extension cords between apartments to pool resources during partial outages. Cafes and restaurants with generators become gathering places during blackouts, offering not just commerce but community and connectivity when home conditions deteriorate.

Psychological Toll and Mental Health

The psychological impact of living under constant threat manifests in ways both obvious and subtle. Acute stress reactions such as panic attacks, hypervigilance and insomnia affect significant portions of urban populations. The distinctive sound of Shahed drones, a buzzing motor noise that Ukrainians learn to recognize instantly, triggers immediate anxiety responses. Many residents report being unable to tolerate similar sounds in daily life, electric shavers, certain types of fans, delivery drones because of associations with attack.

Sleep disruption represents one of the most pervasive and damaging effects. Frequent nighttime alerts fragment sleep cycles, with many Ukrainians averaging only a few hours of continuous rest. The health consequences of chronic sleep deprivation compound over months and years: weakened immune systems, cardiovascular stress, cognitive impairment, and mental health deterioration. Medical professionals observe increases in conditions exacerbated by poor sleep, from hypertension to depression.

Children demonstrate remarkable resilience but also concerning symptoms of trauma. Many Ukrainian children cannot remember life before war, their entire conscious experience shaped by conflict. They play games involving air raid sirens and drone attacks, processing trauma through imagination. Teachers report students having difficulty concentrating, startling at sudden noises, and expressing anxiety about parents and siblings. School psychologists, overwhelmed by case loads, struggle to provide adequate support.

"My daughter is six," shares Natalia, an Odessa accountant. "She was three when the war started. She has vague memories of 'before' but doesn't really remember what normal felt like. For her, this is normal, the sirens, the shelters, the fear. What kind of adult will she become? What psychological scars will she carry? I don't have answers, and it terrifies me."

The mental health system, inadequate before the war, has been overwhelmed by demand. Psychologists and psychiatrists report waiting lists stretching months. Medication for anxiety and depression becomes harder to source as supply chains strain. International organizations provide support, funding therapy services and training additional counselors, but the scale of need vastly exceeds resources. Millions of Ukrainians cope without professional help, relying on family support, community networks, or simply enduring.

Coping mechanisms vary widely. Some residents immerse themselves in work, using productivity as defense against anxiety. Others withdraw, limiting activities to minimize exposure to triggers. Substance use increases, with alcohol consumption rising as people self-medicate against stress. Dark humor pervades Ukrainian social media, residents making jokes about situations that would be unimaginable to outsiders, a psychological defense mechanism that helps maintain sanity amid madness.

The concept of "languishing", a state of reduced wellbeing characterized by emptiness and stagnation without meeting criteria for depression has become widespread. Many Ukrainians describe feeling numb, going through daily motions without joy or purpose, unable to plan for futures that seem impossibly uncertain. This psychological middle ground between thriving and crisis often goes unrecognized and untreated, yet affects quality of life profoundly.

Community Resilience and Adaptation

Despite the hardships, or perhaps because of them, Ukrainian communities have developed remarkable solidarity and mutual support systems. Neighborhood groups organize around shared survival needs, with residents collaborating on shelter maintenance, sharing resources during outages, and maintaining communication networks to warn of approaching threats. Social media groups dedicated to specific districts or buildings provide real-time information about alerts, damage, and safe routes.

Volunteer networks, already extensive from the war's early months, have adapted to serve populations living under drone attack. Organizations deliver supplies to elderly residents unable to navigate stairs during alerts, provide psychological first aid after strikes, and help with debris cleanup and temporary repairs. This volunteer infrastructure, operating largely outside formal government systems, provides crucial support that official agencies cannot match in flexibility and responsiveness.

Cultural life, remarkably, continues. Theaters, museums, and concert halls remain open, hosting performances and exhibitions despite frequent interruptions. Art created during the war reflects the experience of living under attack, with installations about destroyed buildings, performances incorporating air raid sirens, and exhibitions documenting daily life during conflict. Cultural production becomes both processing of trauma and assertion of continued existence, a refusal to let the war completely define Ukrainian identity.

Religious communities play important roles in maintaining morale and providing support. Churches, synagogues, and mosques serve as gathering places, offer shelter during alerts, and provide spiritual comfort to populations facing existential threat. Religious leaders conduct services in basements, deliver sermons addressing the theological questions that war raises, and organize aid distribution. For many Ukrainians, faith provides a framework for understanding suffering and motivation for perseverance.

The concept of "normal" has been redefined rather than abandoned. Ukrainians celebrate birthdays in shelters, hold weddings with contingency plans for alerts, and maintain holiday traditions with modifications for wartime realities. Parks remain popular on quiet days, restaurants serve customers between sirens, and schools conduct graduations despite everything. This insistence on normalcy represents more than denial; its active resistance against allowing the war to completely consume life.

"We refuse to just survive," explains Viktor, a Kharkiv artist who continues exhibiting work despite his city's proximity to the front. "Survival is important, obviously, but we're also determined to live, to create, to celebrate, to plan futures. If we reduce ourselves to mere survival, we've already lost something essential. The Russians can send drones every night, but they cannot take away our humanity unless we surrender it."

Children Growing Up with Drone Sirens

Perhaps no group faces more profound long-term consequences than children growing up during Ukraine's drone war. Developmental psychologists observe that chronic stress during childhood can alter brain development, affect emotional regulation, and shape worldviews in ways that persist throughout life. Ukrainian children are experiencing these formative years amid unprecedented circumstances, with implications that will unfold across decades.

Educational disruption represents one clear impact. Schools nationwide adapted to war conditions, installing shelter spaces, modifying schedules around likely attack times, and developing procedures for rapid evacuation. Yet learning suffers when classes are interrupted multiple times daily, when students arrive exhausted from sleepless nights, and when concentration is impossible amid constant vigilance. Teachers struggle to maintain curricula while addressing students' emotional needs and managing their own trauma.

Online learning, expanded during the war, provides flexibility but creates new challenges. Not all families have reliable internet, electricity, or devices for every child. The isolation of remote learning compounds the social disconnection that children already experience when normal activities of sports, gatherings and celebrations become difficult or impossible. Educational inequality, always present, has widened dramatically as privileged families with resources adapt more successfully than those without.

Play, essential for healthy child development, has been militarized. Children incorporate war themes into games, building forts against imaginary drone attacks, role-playing soldiers and civilians. Toys include miniature drones and military vehicles. Drawings feature explosions, destroyed buildings, and Ukrainian flags. Psychologists debate whether this play represents healthy processing of trauma or normalization of violence that will have long-term consequences. The answer is likely both, varying by individual child and circumstances.

Parenting under these conditions requires navigating impossible dilemmas. How much information should children receive about threats? How does one explain why people are trying to kill them? How to provide security and hope when the future is genuinely uncertain? Parents adopt different approaches, from complete honesty to protective shielding, from maintaining strict routines to flexible accommodation of children's fears. No approach feels adequate; all represent doing one's best in circumstances no one should face.

Some children demonstrate extraordinary resilience, adapting to wartime reality with apparent ease. They excel in school despite disruptions, maintain friendships, and express optimism about futures. Others struggle profoundly, developing anxiety disorders, experiencing nightmares, or withdrawing into themselves. Most fall somewhere between these extremes, showing both strength and vulnerability, coping day-to-day while carrying burdens that will require years to fully process.

"My son asks if we'll still be alive for his birthday next month," shares Maksym, a Zaporizhzhia teacher. "He's eight years old. He shouldn't have to contemplate his own mortality. But he does, because he's seen buildings destroyed, knows classmates who lost family members, and understands that drones can kill. I tell him we'll be fine, that we're safe, but I don't know if that's true. How do you parent honestly while maintaining hope?"

The Normalization of Constant Threat

Three and a half years into the war, what once seemed extraordinary has become ordinary. Air raid sirens that initially triggered panic now prompt irritation at the disruption. Explosions that would have dominated conversation for days are noted briefly before returning to other topics. The human capacity for adaptation, remarkable in its own right, creates its own dangers when adaptation means normalizing persistent threats.

This normalization operates on multiple levels. Physiologically, constant activation of stress responses can lead to dysregulation, where the body no longer appropriately distinguishes between genuine threats and routine situations. Psychologically, the desensitization that allows people to function amid danger can extend inappropriately to other areas, reducing empathy or distorting risk assessment. Socially, when extreme situations persist long enough, they reshape what communities consider acceptable or inevitable.

The ethical implications of normalization trouble many observers. When seven-year-olds discuss drone types with technical precision, when teenagers post shelter selfies as routine social media content, when families debate whether a particular alert warrants sheltering with the casualness of discussing dinner plans, something fundamental has shifted. This shift enables survival but at the cost of innocence, perspective, and perhaps long-term wellbeing.

International attention has waned as Ukraine's war extends. The dramatic footage from 2022, the shocking images of invasion, have given way to grinding attrition that makes less compelling television. Global audiences have moved on to other crises, other concerns. Ukrainians living under attack feel this abandonment keenly, the sense that the world's empathy has expiration dates, that suffering becomes uninteresting when it persists too long.

"We've become boring," says Olena, a Kyiv journalist. "The international media covered us intensively when we were in a new crisis, the breaking story. Now we're old news. Drones hit cities every night, but unless casualties are high or something particularly dramatic happens, it barely registers internationally. We're living through history, experiencing trauma that will mark generations, and the world has mostly stopped watching."

This attention deficit affects not just Ukrainian morale but international support. Aid fatigue sets in among donor nations, military assistance packages face greater political resistance, and the sense of urgency that characterized early responses diminishes. Ukrainians understand that sustaining international engagement requires constantly making the case that their suffering matters, that their struggle deserves continued support, an exhausting additional burden atop everything else.

The Day After: Looking Toward Post-War Recovery

Even as fighting continues, Ukrainians begin contemplating post-war recovery and the challenges that peace will bring. The psychological transition from wartime to peacetime, difficult for any society, will be particularly complex for populations that have lived for years under drone attack. The habits, anxieties, and trauma responses developed during conflict will not disappear when sirens finally stop.

Mental health professionals anticipate massive demand for services as the war ends and survival mode gives way to processing what has been endured. Many Ukrainians currently suppress trauma, focusing on daily necessities and postponing emotional reckoning until safety allows it. When that safety comes, the psychological toll may intensify rather than diminish as people finally have space to feel everything they've survived.

Physical reconstruction will parallel psychological recovery. Destroyed infrastructure must be rebuilt, damaged apartments repaired, scarred cityscapes restored. The scope is staggering, thousands of buildings damaged or destroyed, power generation capacity degraded, transportation networks disrupted. International support will be essential, but the work of rebuilding will take years or decades and require Ukrainian initiative and labor.

The question of how to remember and memorialize the drone war experience troubles many. What monuments should be built? How should civilian suffering be commemorated alongside military sacrifice? Should destroyed buildings be preserved as reminders or removed to enable moving forward? These debates, familiar from other post-conflict societies, will shape how Ukraine processes its trauma and constructs its post-war identity.

Some Ukrainians have already left, seeking safety abroad for themselves and children. This displacement, while understandable, represents loss for Ukraine of talent, youth, and population needed for reconstruction. The country faces the challenge of encouraging return while acknowledging that not everyone will or should come back. Those who remain and those who return will build the post-war society, carrying memories and scars that will influence everything they construct.

"I want my children to know they're Ukrainian, to understand what we survived and why it mattered," reflects Anastasia, a Dnipro entrepreneur planning eventual return from temporary refuge in Poland. "But I also want them to know joy, to experience childhood without constant fear. I don't know how to give them both. Maybe that's the question my entire generation will wrestle with, how to honor what we've endured while not letting it define us completely."

The Testimony of Daily Life

The experience of living under drones, while less dramatic than frontline combat, constitutes its own form of testimony about modern warfare's nature. It reveals how technological capabilities extend conflict into every corner of society, how civilian populations become both targets and front lines, and how the psychological dimensions of warfare can be as devastating as physical destruction.

Ukrainians document their experiences compulsively, posting to social media, maintaining diaries, recording video testimonies. This documentation serves multiple purposes: providing emotional outlet, creating historical record, and asserting to the world that they exist, that their suffering is real, that their stories deserve telling. Future historians will have unprecedented access to daily life during conflict, thanks to this self-documentation.

The ordinariness of most entries makes them powerful. Not dramatic accounts of narrow escapes or heroic resistance, but mundane descriptions of interrupted sleep, cold apartments, anxious children, and relentless fatigue. This is how most people experience war, not through singular moments of clarity or transformation but through grinding accumulation of small hardships that compound into profound disruption.

"People imagine war as dramatic," says Dmytro, a Lviv accountant maintaining a detailed diary. "Explosions, destruction, clear dividing lines between before and after. But mostly it's just exhausting. It's being tired all the time, worried constantly, and trying to maintain routines that feel increasingly absurd. It's checking your phone compulsively for alerts, calculating whether you have enough candles for another blackout, and wondering if you'll ever feel truly safe again."

These testimonies matter not just for Ukraine but for understanding how warfare evolves and what modern conflict means for civilian populations. The lessons from Ukrainian cities under drone attack will inform humanitarian planning, urban defense strategies, and international law governing warfare. More immediately, they provide witness to suffering that deserves acknowledgment, even when the world's attention wanders.

The human capacity for endurance evident in Ukrainian cities is remarkable and troubling in equal measure. Remarkable because people continue living, working, raising children, and maintaining dignity amid circumstances that should be intolerable. Troubling because no one should have to demonstrate such endurance, because adaptation to the unbearable suggests failure by the international community to prevent or stop such suffering.

Life under the drones continues, measured in sleepless nights and interrupted days, in children's questions and parents' inadequate answers, in the buzz of Shaheds overhead and the calculations about whether to shelter or hope for luck. It continues in resilience and trauma, in community solidarity and individual breaking points, in the stubborn insistence on remaining human when everything conspires toward mere survival. This is the civilian face of drone warfare, less visible than battlefield footage but no less important to understanding what modern conflict means and what it costs.

Chapter 29

The Ethics of Drone Warfare

The proliferation of drone warfare in Ukraine has forced urgent confrontation with ethical questions that international law, military doctrine, and philosophical frameworks are ill-equipped to answer. When cheap drones enable precision strikes by anyone with minimal training, when killing becomes as simple as piloting a remote-controlled device, when the barriers to lethal force collapse to near zero, fundamental questions about proportionality, discrimination, accountability, and the very nature of just warfare demand fresh examination. Ukraine's experience, documented in thousands of hours of combat footage and millions of individual targeting decisions, provides unprecedented data for these ethical reckonings while simultaneously revealing how inadequate existing frameworks are for the age of mass drone warfare.

Lowering the Barrier to Lethal Force

Throughout human history, killing in warfare required overcoming significant physical and psychological barriers. Soldiers had to approach enemies close enough to engage with primitive weapons, later with firearms that still demanded proximity and direct confrontation. Even artillery and air power, which introduced distance, required extensive training, expensive equipment, and institutional structures that created multiple decision points between intention and execution. These barriers served functions beyond the merely practical; they provided time for reflection, created accountability through chains of command, and imposed costs that encouraged restraint.

Drone warfare, particularly the FPV systems that have proliferated in Ukraine, strips away most of these barriers. A nineteen-year-old with three weeks of training can kill from kilometers away, watching death unfold through electronic goggles while sitting safely in a dugout. The technical difficulty of the act of flying the drone to target and detonating its payload requires skill but no more than video games demand. The psychological barrier, while not eliminated, is reduced by the mediated nature of the killing, the video-game aesthetic of the interface, and the physical separation from consequences.

This democratization of lethal capability represents a fundamental shift. When sophisticated weapons requiring years of training and millions in equipment were prerequisites for military effectiveness, only states and large organizations could wage serious warfare. FPV drones, assembled from commercial components for hundreds of dollars and operable after brief training, place lethal capability within reach of essentially anyone. The military implications are profound; the ethical implications are staggering.

Critics argue that lowering barriers to lethal force increases violence by making killing easier psychologically and practically. When pulling a trigger required seeing the whites of enemies' eyes, natural human reluctance to kill provided some brake on violence. When pilots dropped bombs from altitude without seeing targets clearly, the abstraction provided psychological distance but also removed visceral understanding of effects. Drone warfare creates a disturbing middle ground: operators see targets with perfect clarity but experience the killing as abstract, separated by screens and distance from the reality of what they do.

Defenders counter that precision and discrimination improve with drones, potentially reducing civilian casualties and unnecessary destruction. A sniper firing blindly at suspected enemy positions might hit civilians; a drone operator can observe targets carefully, confirm military status, and strike with precision that minimizes collateral damage. The argument holds merit but assumes operators use capabilities responsibly, an assumption that combat footage from Ukraine sometimes contradicts.

The question extends beyond individual operators to broader societal implications. If warfare becomes cheaper and easier to conduct, does this make armed conflict more likely? Does the reduced barrier to initiating hostilities destabilize the uneasy peace that fear of war's costs maintains? The evidence from Ukraine suggests affirmative answers. Russia's willingness to conduct sustained drone campaigns against Ukrainian cities reflects calculation that such attacks are affordable and sustainable in ways that traditional air campaigns would not be. Ukraine's ability to strike deep into Russia would be impossible without cheap, proliferated drone technology. Neither side might have undertaken these operations if they required conventional military capabilities.

Proportionality When Drones Are Cheap

International humanitarian law requires that military operations satisfy proportionality: anticipated military advantage must outweigh expected civilian harm. This principle, while subject to interpretation and good-faith disagreement, provides an ethical framework for evaluating battlefield decisions. Drone warfare strains proportionality analysis by radically altering the cost calculations that inform these judgments.

Consider a scenario common in Ukraine: intelligence identifies a Russian command post in a building within a populated area. Under conventional warfare, attacking this target might require aircraft risking pilots' lives and expensive platforms, or artillery expenditure that would deplete limited stocks and reveal firing positions. These costs, human and material factor into proportionality assessments. Commanders must weigh the military value of destroying the command post against risks to attacking forces, expenditure of scarce resources, and potential civilian casualties. The calculation encourages restraint; many targets are simply not worth the cost.

FPV drones transform this calculus. The command post can be struck for five hundred dollars with no risk to Ukrainian personnel. The economic and human costs that previously encouraged restraint essentially disappear. Does this make the attack more proportional because friendly forces face less risk? Or less proportional because the reduced cost lowers the threshold for operations that might harm civilians? The legal framework provides no clear answer.

The abundance of cheap drones creates another proportionality problem: the temptation to strike targets of marginal military value simply because the cost is low. A single Russian soldier taking cover in civilian infrastructure might not warrant artillery or air strike, but a five-hundred-dollar drone? The calculation shifts. Ukrainian forces can and do strike individual soldiers, isolated vehicles, and minor equipment that would never receive attention in conventional warfare. The cumulative effect is militarily significant, death by a thousand cuts grinding down Russian forces but the ethical implications are concerning.

Russia's sustained Shahed campaigns against Ukrainian cities illustrate proportionality collapse from the aggressor's perspective. Each drone costs fifteen to thirty thousand dollars, a fraction of the cost of manned aircraft or cruise missiles. This economic efficiency enables attacks that would be prohibitively expensive with conventional means. Russia launches hundreds of drones nightly against targets of questionable military value, power substations serving primarily civilians, water facilities and residential areas. The low cost per attack enables sustained campaigns that amount to terror bombing, operations that might be restrained by economic considerations if each strike required a million-dollar missile.

The international community has struggled to apply proportionality principles to these drone campaigns. Russia claims legitimate military targets: infrastructure supporting Ukrainian war effort, alleged military facilities in cities, logistics nodes. Ukraine and international observers document strikes on purely civilian targets with no plausible military justification. The abundance and cheapness of drones allows Russia to claim proportionality, minimal military cost for legitimate objectives while actually conducting indiscriminate attacks that proportionality principles were designed to prevent.

Civilian Targeting with Quadcopters

Perhaps the most troubling ethical dimension of Ukraine's drone war involves the use of small quadcopters to directly target civilians and conduct what amounts to individual assassinations. Unlike Shahed drones striking infrastructure or FPV systems targeting military vehicles, these operations involve slow, deliberate stalking of individuals who in many cases are manifestly non-combatant.

Documented incidents include Russian drones following individual civilians in streets, pursuing people attempting to flee, and dropping munitions on clearly marked civilian vehicles. Ukrainian border regions like Kherson, subject to regular drone attacks, have reported drones targeting farmers working fields, families attempting to evacuate, and volunteers delivering humanitarian aid. The operators conducting these attacks can see their targets clearly identifying age, gender, whether they carry weapons or wear uniforms making claims of mistaken identity implausible.

These attacks violate fundamental principles of international humanitarian law. Civilians not taking direct part in hostilities are protected persons who may not be deliberately targeted. The principle of distinction requires belligerents to distinguish between military and civilian targets. Attacks on protected persons constitute war crimes regardless of the weapon employed. That these crimes are committed with quadcopters rather than rifles or artillery changes nothing legally, though the nature of drone operation, watching individuals through cameras while safely distant arguably makes the crimes more calculated and cold-blooded.

Ukraine has documented and publicized these attacks extensively, providing evidence to international criminal investigators and human rights organizations. The footage is damning: drones hovering above civilians who look up at their killers, pursuing fleeing vehicles, striking repeatedly to ensure kills. Russian forces or their proxies conduct these operations with apparent impunity, understanding that the distributed nature of drone warfare makes attribution difficult and prosecution unlikely.

The ethical questions extend to Ukrainian operations as well, though documented instances of Ukrainian forces deliberately targeting civilians are far fewer. Ukrainian drone operators face constant decisions about targets' status: Is that person a combatant or civilian? Is someone in military-age-male demographic automatically a legitimate target? What about individuals who might be conscripts, forcibly mobilized and unenthusiastic participants? Ukrainian doctrine and training emphasize compliance with international law, but enforcement in distributed operations conducted by thousands of individual operators is imperfect.

The personal nature of these attacks distinguishes them ethically from other forms of remote warfare. An artillery crew firing at grid coordinates does not see individuals; a fighter pilot dropping bombs from altitude cannot distinguish faces. Drone operators making targeting decisions through high-definition video feeds see their victims as individuals, observe their final moments, and make conscious choices to kill. This intimate knowledge imposes moral responsibility that cannot be evaded through claims of following orders or acting under uncertainty.

Landmine-Dropping Drones and War Crimes

Among the most egregious uses of drone technology documented in Ukraine involves quadcopters dropping anti-personnel landmines in civilian areas. International humanitarian law, specifically the Ottawa

Treaty (Mine Ban Treaty), prohibits the use, stockpiling, production, and transfer of anti-personnel mines. While neither Russia nor Ukraine is party to this treaty, the deployment of landmines via drone in civilian areas violates customary international law and Geneva Convention provisions protecting civilian populations.

Reports from Ukrainian-controlled Kherson, regularly targeted by Russian drones, describe quadcopters dropping small anti-personnel mines in streets, parks, residential areas, and evacuation routes. These weapons, designed to maim rather than kill, create terror and deny territory to civilians. Children have been injured by mines dropped near schools. Aid workers have been targeted. The indiscriminate nature of landmines, they harm anyone who encounters them, civilian or combatant, long after deployment makes their use in populated areas particularly heinous.

The drone-delivery mechanism adds new dimensions to the landmine problem. Traditional landmine deployment, while criminal, at least required physical access to territories. Minefields could be mapped, marked, and eventually cleared because locations were known or could be reconstructed. Drone-dropped mines in urban environments are scattered randomly, effectively impossible to map comprehensively, and will pose threats to civilians long after hostilities end. The operational advantage these weapons provide is negligible; their purpose appears to be pure terror and civilian harm.

Documentation of these war crimes is extensive. Ukrainian authorities, human rights organizations, and international investigators have collected physical evidence, interviewed victims, and analyzed drone footage showing mine deployment. The evidence is sufficient for criminal prosecution, assuming perpetrators can be identified and brought to justice. The challenge lies in attribution, determining which units and individuals conducted specific attacks and in the political will to pursue accountability.

The legal framework for prosecuting drone-enabled war crimes remains underdeveloped. Existing international criminal law addresses murder, targeting civilians, and weapons violations, but application to distributed drone operations conducted by hundreds of individual operators poses practical challenges. How does one identify the specific individual who piloted a particular drone? How to establish command responsibility when operations are decentralized? How to prosecute crimes conducted via technology that leaves minimal physical evidence beyond the effects?

These questions will occupy international lawyers and courts for years. The International Criminal Court has opened investigations into crimes in Ukraine, including drone-related offenses. Ukrainian authorities conduct their own investigations and have vowed prosecution of Russian personnel identified through captured equipment, communications intercepts, and other evidence. Whether these efforts ultimately deliver justice remains uncertain, but the documentation being compiled ensures that evidence exists for future accountability.

The Documentation of Drone Footage

An unusual feature of Ukraine's drone war is the extensive documentation of combat operations through onboard cameras. Most FPV drones record their flights, creating permanent visual records of strikes. This footage serves operational purposes, confirming kills, analyzing tactics and training operators but also creates an unprecedented historical record of warfare's intimate details. The ethical implications of this documentation merit consideration.

The footage depicts death with clarity and detail rarely available in previous conflicts. Viewers see soldiers' final moments, watch as vehicles explode with occupants inside, observe the effects of explosive detonations on human bodies. This documentation removes abstraction from warfare, forcing confrontation with violence's reality. For some viewers, this exposure provides important context for understanding war's true nature. For others, it becomes pornography of violence, sensationalized content consumed for entertainment rather than education.

Ukrainian authorities have made strategic decisions to release certain footage publicly, using documentation to demonstrate military effectiveness, prove Russian atrocities, and maintain international support. These releases require careful curation, what to show, what to withhold, how to present material that informs without gratuitously depicting suffering. The balance is difficult. Too much sanitization and audiences miss warfare's reality; too much graphic content and audiences become desensitized or disgusted.

The soldiers who appear in this footage, particularly Russian personnel killed in documented strikes, raise ethical questions about dignity and respect for the dead. International humanitarian law requires that deceased combatants be treated with respect, their remains handled appropriately, and their identities protected. Does broadcasting someone's death violate these principles? Does it matter that the footage serves legitimate military and historical purposes? Is there a meaningful difference between documenting combat deaths and exploiting them?

Consent and privacy, concepts that barely applied to battlefield recording previously, become relevant when every engagement generates video evidence that might be viewed by millions. Russian soldiers presumably did not consent to their deaths being filmed and broadcast globally. Ukrainian soldiers whose operations are recorded might not wish their actions scrutinized by audiences who were not present and cannot understand combat's context. The permanent nature of digital documentation means these recordings will exist indefinitely, accessible to future generations who will judge past actions by their contemporary standards.

Social media platforms hosting drone combat footage face content moderation dilemmas. This material violates policies against graphic violence and could traumatize viewers, particularly children who access platforms despite age restrictions. Yet the footage also documents historically important events and war crimes that deserve international attention. Platform policies attempting to balance these concerns satisfy no one fully. Ukrainian advocates argue that restricting combat footage aids Russian propaganda by hiding the war's reality. Critics argue that platforms have become distribution channels for violence that numbs audiences and glorifies killing.

Accountability in Distributed Warfare

Traditional military operations feature clear chains of command, explicit orders, and institutional structures that create accountability. An artillery strike requires multiple authorizations, coordination with higher headquarters, and documentation that establishes who ordered the attack and why. Air operations involve even more complex approval processes. These systems exist partly for operational coordination but also to ensure accountability, so that responsibility for controversial or potentially criminal actions can be traced to individuals.

Drone warfare in Ukraine often operates with far less structure. Individual operators or small teams identify targets, make engagement decisions, and execute attacks with minimal oversight. This decentralization enables agility and responsiveness but creates accountability problems. When attacks go wrong, when civilians are killed, when war crimes occur establishing responsibility becomes difficult.

Who bears responsibility when a nineteen-year-old drone operator kills a civilian he mistakenly identified as combatant? The operator himself? His immediate commander who provided minimal supervision? Higher officers who created distributed operational structures? Political leaders who authorized aggressive drone operations? Legal frameworks suggest responsibility should be shared among these parties according to their roles, but practical accountability often fails to materialize.

The scale of drone operations compounds accountability challenges. When thousands of drones fly daily, operated by hundreds of pilots, conducting thousands of strikes, comprehensive oversight becomes impossible. Commanders cannot review every targeting decision in real time. Post-operation reviews can examine only a small fraction of missions. Quality control relies on operator judgment and peer accountability within units' culture rather than institutional safeguards that may prove inadequate.

Ukrainian military authorities recognize these challenges and have implemented measures to improve accountability. Rules of engagement specify legitimate targets and prohibited actions. Training emphasizes international humanitarian law compliance. Units are supposed to maintain flight logs documenting missions, targets, and outcomes. Yet enforcement is imperfect, particularly for volunteer units operating with minimal official oversight and for operations conducted under intense combat pressure.

Russian accountability appears virtually nonexistent for drone operations targeting Ukrainian civilians. Operators conducting attacks on protected persons face no apparent disciplinary action, suggesting explicit authorization or at minimum willful blindness by commanders. The command structure authorizing terror bombing campaigns against Ukrainian cities is well-established, with responsibility traceable to senior military and political leadership. Whether this leadership will ever face justice remains a question of international will rather than legal ambiguity.

The distributed nature of drone warfare creates attribution problems that may enable impunity. Unlike conventional operations where units can be identified through equipment signatures, radio communications, and logistics trails, drone attacks often leave minimal evidence. A small quadcopter dropping a munition or an FPV strike might be conducted by any of hundreds of units operating in a sector. Without distinctive markings, telemetry data, or captured equipment, determining which specific unit and operators conducted particular attacks proves difficult.

Philosophical Questions Without Answers

Beyond legal frameworks and institutional accountability lie deeper philosophical questions about drone warfare that military ethicists, political theorists, and moral philosophers are only beginning to address. These questions concern the nature of courage, the meaning of sacrifice, the relationship between risk and moral authority, and whether forms of warfare that minimize friendly casualties while maximizing enemy deaths can ever satisfy just war principles.

Traditional military ethics valorize courage, sacrifice, and warriors' willingness to risk death in defense of legitimate causes. This risk-acceptance provides moral weight to military service and limits warfare's scope. Societies are restrained in initiating conflicts when their children's lives are at stake. Drone warfare conducted from safe positions undermines this moral economy. If killing becomes risk-free for one side, does this change warfare's moral status? Is there something troubling about killing enemies while accepting no reciprocal vulnerability?

Some philosophers argue that asymmetric risk exposure is inherently problematic, that just warfare requires some equality of risk among belligerents. By this logic, drone operators killing from safety engage in something less than legitimate combat, more like execution than warfare. Others counter that soldiers have no obligation to accept unnecessary risk, that protecting one's forces while degrading enemy capabilities is a fundamental military objective, and that moral legitimacy derives from cause rather than mutual vulnerability.

The question of moral injury revisited in philosophical context reveals tensions in drone warfare's ethical status. Operators report psychological distress despite physical safety, suggesting that something about remote killing troubles human conscience. Is this distress evidence that drone warfare violates moral intuitions that have evolved over millennia of human conflict? Or is it simply adjustment to new technology that future generations will regard as normal as we regard artillery or air power?

The relationship between technological capability and moral restraint poses perhaps the most troubling question. Throughout history, technological limitations imposed some restraint on violence. When killing required physical proximity, natural human reluctance provided a brake on bloodshed. When weapons were expensive, economic considerations limited their use. When operations risked friendly casualties, concern for soldiers' lives encouraged restraint. Drone warfare removes many of these practical limits. If the only constraints on violence are legal rules and moral conviction unsupported by practical barriers, can we trust these will prove sufficient?

Ukraine provides no definitive answers to these questions, only urgent reminders of their importance. The ethical frameworks developed for previous forms of warfare strain under drone proliferation's pressure. New principles adequate to drone warfare's reality have not emerged. In this gap between obsolete ethics and unrealized new frameworks, operators make individual moral choices, commanders navigate unprecedented dilemmas, and societies grapple with warfare's transformation. The ethics of drone warfare remain contested territory, with more questions than answers and stakes that could not be higher.

PART VII: THE FUTURE

Chapter 30

What Comes Next

The drone warfare revolution witnessed in Ukraine between 2022 and 2025 represents not a culmination but an opening chapter. The systems, tactics, and production scales that seem extraordinary today will appear primitive within a decade. The strategic calculations, ethical dilemmas, and military adaptations that Ukraine has forced upon the world are merely initial responses to transformations that will continue accelerating. Understanding where drone warfare is headed requires examining technological trajectories already visible, extrapolating from current trends, and grappling with possibilities that sound like science fiction but rest on engineering foundations being laid today.

Continued Escalation in Numbers and Sophistication

Ukraine's achievement of 2.5 million annual production capacity with plans to reach 4 million represents remarkable scaling, but these numbers will not mark ceilings for long. The industrial processes, supply chains, and expertise developed during the conflict create foundations for even more dramatic expansion. A nation that built drone production from essentially nothing in three years could plausibly double or triple capacity again in the next three years given sufficient motivation and resources.

The economics driving this scaling remain compelling. As manufacturing processes mature, unit costs decline through economies of scale, learning effects, and supply chain optimization. What cost five hundred dollars to produce in 2023 costs three hundred in 2025 and might cost two hundred by 2027. Simultaneously, operational experience refines designs toward optimal cost-performance tradeoffs, eliminating unnecessary capabilities that add expense without proportional benefit. The trajectory points toward even cheaper systems produced in even larger quantities.

Other nations observing Ukraine's experience are beginning their own scaling efforts. China, already the world's dominant commercial drone manufacturer, possesses industrial capacity that dwarfs Ukraine's. If

Chinese military procurement shifted even a fraction of commercial production toward military systems, output could reach tens of millions annually. Russia, having established Shahed production using smuggled components and Iranian technology transfer, continues expanding capacity. Western nations, slower to adapt but finally recognizing strategic necessity, are initiating procurement programs measured in hundreds of thousands of units.

The global trajectory suggests that within a decade, major powers will maintain standing inventories of millions of drones, with production capacity to sustain attrition rates orders of magnitude higher than anything seen in Ukraine. This abundance will fundamentally alter military planning. When drones are so numerous that losing hundreds daily is sustainable, their employment will become even more profligate. Saturation attacks that today involve hundreds of drones will involve thousands. The aerial battlefield will become unimaginably dense and lethal.

Sophistication will advance alongside quantity. Current Ukrainian and Russian drones, while effective, remain relatively simple systems optimized for low-cost mass production. Future iterations will incorporate more advanced sensors, improved guidance systems, longer range, and greater payload capacity while maintaining affordability. The integration of artificial intelligence, still in early stages, will enable capabilities currently impossible. The drones of 2030 will bear the same relationship to current systems that modern FPV drones bear to the commercial quadcopters of 2022, recognizably in the same family but dramatically more capable.

The AI Arms Race

Artificial intelligence represents the next frontier in drone warfare, promising capabilities that will make current systems appear crude by comparison. Ukrainian manufacturers and research institutions are already developing AI-enabled autonomous systems, though operational deployment remains limited by technical challenges, regulatory concerns, and ethical considerations. These limitations will erode rapidly as technology matures and competitive pressures intensify.

Machine vision and target recognition constitute the most immediate AI applications. Current FPV drones require human operators to identify targets, navigate to them, and time detonation. Tasks demanding skill, concentration, and continuous connectivity. AI systems can automate these functions, enabling drones to identify military vehicles, distinguish combatants from civilians, navigate to targets autonomously, and detonate at optimal moments. This automation dramatically reduces operator workload, enables each operator to control multiple drones simultaneously, and functions in GPS-denied or communications-jammed environments.

Ukrainian developers have demonstrated prototype systems with these capabilities in controlled environments. Drones equipped with machine vision can identify and track specific vehicle types, Russian tanks versus Ukrainian ones, military trucks versus civilian vehicles. The systems achieve accuracy rates exceeding ninety percent in clear conditions, though performance degrades with poor visibility, camouflage, or deliberate deception. Current limitations include high computational requirements, vulnerability to adversarial techniques that fool recognition algorithms, and reliability concerns that make military operators reluctant to trust automated targeting.

These limitations are temporary. Processing power increases exponentially according to Moore's Law trends, making sophisticated AI implementable on progressively smaller, cheaper hardware. Training data from Ukraine's conflict, millions of hours of drone footage documenting every conceivable tactical scenario provides unprecedented resources for refining recognition algorithms. Adversarial robustness improves through techniques that expose AI systems to deception attempts during training. Within five years,

AI-enabled autonomous targeting will likely achieve reliability that exceeds human operators in many scenarios.

Swarming technologies represent AI's most transformative application in drone warfare. Current operations involve individual drones controlled by individual operators, enabling at most a few simultaneous strikes. AI-enabled swarms could coordinate dozens or hundreds of drones executing complex, synchronized operations. A swarm might saturate air defenses from multiple vectors simultaneously, adapt tactics in real-time based on enemy responses, and concentrate effects on high-value targets identified through collective reconnaissance.

The military advantages of swarming are profound. Coordinated swarms overwhelm defenses far more effectively than equivalent numbers of individually controlled drones. They enable tactical complexity impossible for human operators to manage. They maintain effectiveness despite heavy attrition as remaining drones redistribute roles autonomously. The side that fields effective swarms first will possess enormous advantages over adversaries relying on individually piloted systems.

Research into drone swarming is intensive globally. Chinese military research institutes publish papers on swarm algorithms and demonstrate prototype systems. The U.S. Department of Defense funds multiple swarm programs exploring different technical approaches. Israeli companies develop commercial systems with dual-use implications. Ukraine's wartime experience provides unmatched testing data but may lack resources for cutting-edge AI research. The nation that achieves breakthrough in militarily effective swarming will gain strategic advantages that could prove decisive in future conflicts.

Autonomous navigation in GPS-denied environments addresses one of drone warfare's current vulnerabilities. Electronic warfare and GPS spoofing currently degrade drone effectiveness significantly. Fiber-optic guidance provides one solution but imposes range limitations. AI-enabled visual navigation using cameras and inertial sensors to navigate by terrain features and landmarks offers an alternative that combines immunity to jamming with long-range capability. Military significance is obvious: drones that cannot be jammed or spoofed operate with impunity in contested electromagnetic environments.

The integration of these AI capabilities will proceed unevenly. Simple automation will deploy first: automated target tracking, trajectory optimization, obstacle avoidance. More sophisticated capabilities requiring greater computational resources and algorithmic advances will follow. Full autonomy, where drones select targets and prosecute attacks without human intervention, represents the ultimate capability and the most concerning ethically. This progression from human-in-the-loop to human-on-the-loop to fully autonomous systems is technologically inevitable and strategically irresistible, though its ethical and legal implications remain deeply contested.

Hypersonic Drones on the Horizon

The next category jump in drone capabilities involves hypersonic systems capable of speeds exceeding Mach 5. Current drones, whether FPV quadcopters or Shahed-type platforms, fly at speeds comparable to automobiles or small aircraft, fast enough to make interception challenging but slow enough that capable air defenses succeed regularly. Hypersonic drones would compress decision timelines to seconds, rendering current defensive systems largely ineffective.

Hypersonic technology currently exists primarily in expensive, sophisticated missile systems developed by major powers. China's DF-17, Russia's Kinzhal and Avangard, and U.S. developmental programs represent state-of-the-art in hypersonic weapons. These systems cost millions of dollars each and require advanced materials, precision manufacturing, and complex guidance systems. They are not drones in the Ukrainian sense, cheap, numerous, expendable platforms but rather strategic weapons comparable to ballistic missiles.

However, the engineering principles enabling hypersonic flight are not fundamentally exclusive to expensive systems. As materials science advances, manufacturing techniques improve, and production scales increase, the cost barriers to hypersonic capabilities will decline. Already, research programs worldwide explore low-cost hypersonic designs using simplified scramjet engines, expendable airframes, and minimalist guidance. The goal is hypersonic platforms costing tens or hundreds of thousands of dollars rather than millions, still expensive by FPV standards but affordable enough for mass production.

The strategic implications of hypersonic drones are staggering. Current air defenses intercept ninety-three percent of subsonic drones through combination of detection, tracking, and kinetic or electronic engagement. Against hypersonic targets, interception becomes vastly more difficult. Detection time compresses, tracking requires faster systems, and kinetic intercepts demand perfect timing and positioning. A ninety-three percent success rate against subsonic drones might translate to thirty percent against hypersonics, a defense penetration rate that makes saturation attacks devastatingly effective.

The timeline for affordable hypersonic drones remains uncertain but probably measured in years rather than decades. Prototype systems exist in laboratories. Engineering challenges, while substantial, are not insurmountable. The nation that achieves breakthroughs to produce hypersonic drones will gain temporary but decisive strategic advantage, a first-mover benefit similar to what nuclear weapons provided the United States in 1945. This incentivizes aggressive research spending and risk-taking in development programs across multiple nations.

Ukraine, despite its impressive drone capabilities, likely lacks the resources and industrial base for hypersonic development without substantial international partnership. Russia has hypersonic expertise but faces sanctions constraining access to necessary components. China possesses both the technology base and manufacturing capacity for rapid progress. Western powers, particularly the United States, have technological advantages but may move more slowly due to bureaucratic procurement processes. The hypersonic drone race will shape strategic balance for decades.

Smaller, Smarter, Deadlier

While some technological trajectories point toward larger, faster, more sophisticated systems, others move in the opposite direction: toward increasingly miniaturized drones with capabilities packed into ever-smaller form factors. The physics of small systems, lower energy requirements, reduced acoustic and electromagnetic signatures, difficulty of detection provides significant military advantages that current drone designs have barely begun exploiting.

Commercial micro-drones already exist with dimensions measured in centimeters and weights under fifty grams. These systems, marketed for indoor photography and entertainment, demonstrate feasibility of highly miniaturized flight platforms. Military applications would add modest payloads, perhaps fifty grams of high explosive, enough to kill or severely injure individuals and guidance systems. The resulting micro-drones would be nearly impossible to detect acoustically, difficult to track visually, and essentially immune to current counter-drone systems designed for larger targets.

The tactical implications are profound and disturbing. Micro-drones could infiltrate buildings through windows or ventilation systems, navigate corridors autonomously, identify targets through AI-enabled recognition, and strike with precision. They could loiter undetected in urban environments, activating only when high-value targets appear. They could be deployed in swarms of thousands, creating aerial clouds of lethal autonomous weapons that transform battlespaces entirely.

Current counter-drone systems rely on detecting electronic emissions, acoustic signatures, or visual identification. Micro-drones could be designed to minimize all these signatures below detection thresholds. Electronic emissions can be reduced through optical or inertial navigation. Acoustic signatures diminish with size. Visual detection of centimeter-scale objects flying at speed approaches impossibility. The defensive challenge micro-drones pose has no obvious solution with current technology.

International discussions about autonomous weapons and "killer robots" often focus on hypothetical future systems. Micro-drones represent near-term realization of these concerns. The technology for autonomous micro-drones exists today; only integration, refinement, and production scaling separates concept from reality. Some estimates suggest widespread deployment could occur within five years absent international agreements constraining development.

The ethical concerns about miniaturized autonomous weapons are severe. The lowered barrier to lethal force reaches its nadir when weapons become so small, cheap, and easy to deploy that their use approaches casualness. The potential for abuse by state and non-state actors is obvious. The difficulty of defense against micro-drone swarms creates vulnerability that even powerful nations struggle to address. The future that micro-drones portend is one where personal security and physical safety erode further, where asymmetric capabilities enable small groups to threaten large populations, and where the line between warfare and terrorism blurs into irrelevance.

Counter-Drone Evolution

As offensive drone capabilities advance, counter-drone technologies must evolve correspondingly or defenses will collapse entirely. Current systems achieving ninety-three percent interception rates against subsonic drones represent impressive engineering but cannot be extrapolated indefinitely. When attack volumes increase tenfold, when hypersonic drones compress reaction times, when micro-drones operate below detection thresholds, fundamentally new defensive approaches become necessary.

Directed energy weapons such as lasers and high-powered microwaves represent the most promising counter-drone technology for dealing with mass attacks. Unlike kinetic systems that require physical interceptors for each target, directed energy weapons can engage multiple targets rapidly with minimal cost per shot. A laser system capable of burning out drone electronics or destroying control surfaces can engage dozens of drones per minute at trivial marginal cost. High-powered microwave weapons can disable multiple drones simultaneously across wide areas.

The United States, Israel, China, and other nations have developed prototype directed energy counter-drone systems with demonstrated effectiveness. Current limitations include power requirements, atmospheric effects that degrade beam quality, and relatively short effective ranges. These limitations will diminish as technology matures. Compact, high-power electrical storage enables mobile deployment. Adaptive optics compensate for atmospheric distortion. Multiple emitters can be networked for overlapping coverage.

Within a decade, directed energy weapons will likely be standard components of military air defense systems. Forward bases will mount laser systems providing continuous coverage against drone attacks. Naval vessels will integrate directed energy weapons protecting against swarms. Even ground vehicles might carry miniaturized systems providing local defense. The transition from kinetic to directed energy interception will enable sustainable defense against attack volumes that would overwhelm conventional systems.

Electronic warfare will continue expanding in sophistication and importance. Current jamming and GPS spoofing provide partial defense but can be countered through frequency hopping, inertial navigation, and fiber-optic guidance. Next-generation electronic warfare will employ cognitive radio techniques that identify and attack communication protocols adaptively, directed electromagnetic pulses that disable electronics within confined areas, and AI-driven systems that predict and preempt drone tactics.

Drone-on-drone combat represents another evolutionary path. Rather than relying on ground-based systems to intercept aerial threats, defender drones could engage attacker drones in aerial combat. Small, agile interceptor drones equipped with nets, projectiles, or electronic warfare packages could patrol likely approach routes, destroy incoming drones before they reach targets, and do so at costs comparable to threats they counter. This approach essentially extends the aerial battlefield, creating contested airspace where drone combat occurs constantly.

The counter-drone technology race will shape military effectiveness as significantly as offensive capabilities. Nations that develop superior counter-drone systems will negate adversaries' numerical advantages. Those that fall behind will find their expensive infrastructure vulnerable to cheap attacks. The asymmetry between offense and defense in drone warfare currently favors offense heavily; future technological developments might restore some balance or might increase offensive advantages further.

The Industrialization of Warfare

Perhaps the most profound implication of drone warfare's evolution is the complete industrialization of combat is the transformation of military effectiveness from primarily a function of training, courage, and tactical skill into primarily a function of industrial production capacity. This shift, already evident in Ukraine, will intensify as drone warfare matures and spreads.

Traditional military power rested on professional armed forces: soldiers who trained for years, pilots who accumulated thousands of flight hours, crews who maintained complex equipment. These human capabilities, combined with sophisticated weapons platforms, determined military effectiveness. Industrial capacity mattered but primarily in providing equipment for professionals to employ. A nation with superior training and doctrine could defeat numerically superior but less capable adversaries.

Drone warfare inverts this relationship. When systems are cheap enough to be expended freely, simple enough that minimally trained operators achieve effectiveness, and numerous enough that quantity itself becomes quality, industrial capacity becomes the primary determinant of military power. The side that can produce and field more drones wins through attrition and saturation regardless of individual system quality or operator skill. Ukraine and Russia both learned this lesson; future conflicts will reflect it from inception.

This shift disadvantages nations that traditionally relied on qualitative advantages. Western militaries emphasizing professionalism, advanced technology, and superior training will find these advantages eroded when adversaries field drone swarms in quantities that overwhelm defenses. The United States, for example, maintains military dominance partly through technological superiority and highly trained forces. Against adversaries producing millions of drones annually, these advantages diminish significantly.

Conversely, the industrialization of warfare advantages nations with large manufacturing bases and capacity for rapid scaling. China's industrial might, already formidable, becomes a strategic weapon of first importance when military effectiveness correlates directly with production volumes. Russia's willingness to mobilize industry for military production, combined with access to Iranian technology and Chinese components, enables sustained operations despite economic weakness by conventional measures.

For smaller nations, the industrialization of warfare presents both opportunities and threats. The barrier to developing credible military capabilities lowers when effectiveness depends on production rather than expensive training and platforms. Nations that develop drone manufacturing can achieve deterrence previously impossible. Yet these same nations face vulnerability to larger adversaries' industrial might if conflicts become attritional.

The policy implications extend beyond military procurement into national industrial strategy, supply chain security, and economic planning. Nations must maintain or develop manufacturing capabilities for drone production, secure access to critical components, and build capacity for rapid production scaling in crisis. The peacetime economy and wartime capability become inseparable. Every nation must consider how quickly it could transition commercial manufacturing to military production, a calculation that recalls Total War planning of twentieth century conflicts but adapted to twenty-first century technology.

Ukraine demonstrated that rapid industrial mobilization remains possible for nations with engineering talent, manufacturing infrastructure, and existential motivation. What Ukraine achieved in three years provides a template for others but also reveals how disruptive such mobilization is and how much it depends on particular circumstances. The question facing every nation is not whether to prepare for industrialized drone warfare but how quickly and completely to commit resources to this preparation versus competing priorities.

The Acceleration of Everything

Underlying all these technological trajectories is an acceleration of change itself. The pace of innovation in drone warfare exceeds anything seen in military technology since World War II's extraordinary compression of development cycles. What took years in peacetime requires months in war; what took months now requires weeks. This acceleration shows no signs of slowing and may intensify as AI enables even faster iteration and as competitive pressures mount.

Ukrainian manufacturers routinely field new drone variants within weeks of identifying requirements. Design modifications happen daily based on operator feedback. Tactical innovations spread through forces in days via social media and informal networks. Russian systems adapt quickly to Ukrainian countermeasures and vice versa. The OODA loop of observe, orient, decide and act that military theorists emphasized operates at speeds that doctrinal development cannot match.

This acceleration advantages agile organizations capable of rapid iteration but disadvantages bureaucratic institutions requiring approval layers and formal processes. Military organizations worldwide struggle adapting to this pace. The gap between innovative combat units and sclerotic procurement systems widens constantly. The question is whether traditional military institutions can reform sufficiently to match the pace that technology and wartime pressure demand or whether new organizational forms will prove necessary.

For defense planning and strategic forecasting, the acceleration creates profound uncertainty. Capabilities considered decisive today may be obsolete tomorrow. Investments in particular technologies or platforms risk being overtaken by developments before deployment. The traditional approach of planning for conflicts years in advance becomes questionable when technology evolves faster than planning cycles. Militaries must somehow balance commitment to long-term programs with flexibility to adapt rapidly to unforeseen changes, a balance that no one has fully achieved.

The future of drone warfare is not predetermined but neither is it uncertain. The technological trajectories are clear, the industrial imperatives established, the strategic logic compelling. What comes next is more drones, smarter drones, deadlier drones, and militaries completely restructured around drone warfare as the primary operational paradigm. The revolution that Ukraine initiated will continue accelerating, transforming warfare more completely in the next decade than in the previous century. The only certainty is that change will continue, accelerate, and surprise.

Chapter 31

Global Proliferation

The technologies, tactics, and industrial processes demonstrated in Ukraine are not remaining contained within that conflict. They are spreading rapidly across the globe, diffusing through multiple channels to state militaries, non-state actors, insurgent groups, and terrorist organizations. This proliferation represents one of the most significant security challenges of the coming decades, fundamentally altering threat landscapes in regions far removed from Eastern Europe. The genie that Ukraine and Russia released from the bottle cannot be returned, and its spread will reshape conflicts worldwide in ways both predictable and terrifying.

Non-State Actors Watching and Learning

Insurgent groups, militant factions, and terrorist organizations have observed Ukraine's drone war with intense interest, recognizing that capabilities once requiring state-level resources are now accessible to any group with modest funding and technical expertise. The implications for asymmetric warfare, counter-insurgency operations, and global terrorism are profound and deeply troubling.

The tactical lessons from Ukraine translate directly to non-state actor needs. FPV drones provide precision strike capabilities against vehicles, fortified positions, and personnel without requiring sophisticated training or expensive platforms. Reconnaissance drones enable surveillance that ground observers cannot achieve. Long-range strike drones allow attacks on targets previously beyond reach. All of these capabilities are implementable with commercial components and publicly available designs derived from Ukrainian innovations.

Middle Eastern militant groups have already begun incorporating drone capabilities learned from Ukraine's experience. Hezbollah, which previously employed relatively simple Iranian-supplied drones, is reportedly developing more sophisticated systems based on FPV concepts demonstrated in Ukraine. Houthi forces in Yemen, already proficient with Iranian drone technology, are adapting Ukrainian tactical innovations to their conflict against Saudi Arabia and maritime targets in the Red Sea. Hamas and Palestinian Islamic Jihad have shown increasing drone sophistication in operations against Israel, with designs bearing unmistakable influence from Ukrainian systems.

The Syrian civil war, which already featured significant drone employment by multiple factions, is seeing rapid evolution as fighters return from Ukraine bringing expertise and equipment. Syrian opposition groups have access to Ukrainian volunteers who fought against Russia and now share knowledge with groups opposing Assad's regime and its Russian backers. The technical cross-pollination happening in Syrian conflict zones will accelerate drone proliferation throughout the Middle East.

African insurgencies are beginning to incorporate drone capabilities as well. Groups operating in the Sahel region, in Somalia, and in other conflict zones have demonstrated increasing interest in unmanned systems. While most currently lack the technical sophistication for advanced systems, the barriers are lowering constantly. Nigerian groups, Somali militants, and various factions in Central African conflicts will likely field meaningful drone capabilities within years if not months.

The financial barriers to drone warfare are low enough that essentially any organized group can afford basic capabilities. An FPV drone capable of destroying a vehicle costs less than the rifles many insurgent fighters carry. Reconnaissance quadcopters cost no more than cell phones. The knowledge required is disseminating through online communities, instruction videos, and diaspora networks. Former Ukrainian volunteers, both foreign fighters returning home and refugees displaced by conflict, carry expertise that spreads through personal connections.

The strategic implications for counter-terrorism and stability operations are severe. Western military forces that struggled against improvised explosive devices in Iraq and Afghanistan will face enemies equipped with precision strike capabilities from air. Small outposts, patrol routes, and logistics convoys that were vulnerable to ambush will become even more exposed. The technological and training advantages that conventional militaries enjoy will erode as adversaries acquire capabilities that negate some traditional strengths.

The Drone Technology Diffusion

The mechanisms through which drone technology spreads deserve examination because they operate differently from traditional weapons proliferation. Nuclear weapons require specialized materials, advanced manufacturing, and extensive technical expertise that create bottlenecks enabling some control. Drone technology faces none of these constraints, spreading through commercial supply chains, open-source designs, and knowledge transfer that is nearly impossible to interdict.

Commercial availability of components represents the primary proliferation pathway. The motors, batteries, flight controllers, cameras, and other subsystems that comprise military drones are identical to those in commercial systems. Manufacturers in China, Taiwan, and other countries produce these components by the millions for legitimate markets. Any purchaser can acquire them without restrictions or special licensing. Assembly requires no specialized equipment beyond basic electronics tools available at consumer retailers.

The open-source nature of much drone technology accelerates diffusion. Enthusiast communities have developed sophisticated flight control software, ground station applications, and manufacturing techniques that they share freely online. Forums, GitHub repositories, YouTube channels, and social media groups provide comprehensive instructions for building capable drones. Much of this information was generated innocuously by hobbyists but proves directly applicable to military systems.

Ukrainian innovations specifically are spreading through multiple channels. Foreign volunteers who fought in Ukraine return home with firsthand knowledge of systems and tactics. Ukrainian refugees include engineers and technicians who worked in drone manufacturing and carry expertise to destination countries. Social media posts, news coverage, and documentary content provide detailed technical information. Ukrainian manufacturers seeking post-war export markets share specifications and designs with potential customers globally.

The knowledge transfer cannot be stopped by export controls or technology restrictions because the knowledge itself is not classified and cannot be made classified retroactively. Attempts to restrict information sharing run into free speech protections in democratic societies and prove unenforceable in authoritarian ones. The distributed nature of knowledge storage, millions of copies on personal devices, cloud services, and distributed networks makes comprehensive suppression impossible even if it were legally feasible.

3D printing technology provides another proliferation vector by enabling local manufacturing of components that previously required specialized facilities. While complete drones cannot yet be 3D printed practically, many components can be produced with consumer-grade printers. Airframes, mounting brackets, protective housings, and various structural elements can be fabricated locally. As 3D printing technology advances, the portion of a drone that can be printed locally will increase, further reducing dependence on controlled supply chains.

The dual-use nature of drone technology makes effective control essentially impossible. Governments can restrict sales of complete military drones, but they cannot restrict components also used in agriculture, infrastructure inspection, delivery services, photography, and countless legitimate applications. Attempting to control component sales would damage enormous legitimate industries while achieving minimal security benefit. Criminals and terrorists would simply manufacture or acquire components through black markets.

Low Barriers to Entry

The accessibility of drone technology creates threats from adversaries that previously lacked capability for sophisticated attacks. A determined individual with a few thousand dollars and several months of self-study can build systems capable of causing significant damage. Small groups with modest resources can field dozens or hundreds of drones. Organizations with access to millions in funding can scale to capabilities approaching those of small state militaries.

The technical barriers have been lowered to the point that high school students regularly build capable drones as science fair projects. University engineering programs include drone design and construction as standard coursework. Countless online tutorials walk viewers through assembly step-by-step. The learning curve that once took years can now be compressed into weeks or months for motivated individuals.

The operational barriers are equally low. Flying FPV drones requires practice but not extensive training. Most operators achieve basic competence within hours and mission-ready capability within days or weeks. The skills are similar to video gaming, which millions of young people worldwide have developed through entertainment. The generation raised on first-person shooter games possesses intuitive capabilities for operating FPV drones that older generations lack.

Targeting and mission planning, once specialized military skills, are facilitated by commercial software and smartphone applications. Mapping software provides detailed imagery and terrain data. Flight planning tools optimize routes and calculate fuel requirements. Targeting can be refined through publicly available satellite imagery and open-source intelligence techniques. The operational planning that previously required trained staff can now be conducted by individuals using consumer electronics.

The sustainability of drone operations poses minimal logistical challenges. Systems are reusable until destroyed, requiring only battery charging between missions. Maintenance requires basic tools and skills comparable to bicycle repair. Spare parts are commercially available or can be cannibalized from other drones. Groups can maintain operational capability indefinitely without sophisticated supply chains or technical support infrastructure.

This accessibility transforms conflict dynamics fundamentally. Whereas insurgent groups previously relied on small arms, improvised explosives, and crude rockets, they now can field precision strike capabilities and persistent surveillance. Terrorist organizations can conduct attacks with drones carrying explosives, biological agents, or radiological materials. Criminal enterprises can use drones for smuggling, surveillance of law enforcement, and targeted assassination.

Terrorism Implications

The application of proliferated drone technology to terrorism represents among the most concerning proliferation scenarios. Terrorist groups have long sought capabilities for spectacular attacks causing mass casualties and generating international attention. Drones provide multiple pathways toward these objectives while complicating defensive responses.

The most straightforward terrorist application involves explosive-laden drones targeting crowds, infrastructure, or symbolic locations. A quadcopter carrying several kilograms of explosives could strike a stadium during a sporting event, a concert venue during a performance, or a transportation hub during rush hour. The attack would be difficult to prevent, nearly impossible to defend against with current security measures, and would generate enormous psychological impact even if physical casualties were moderate.

Biological or chemical weapons delivered by drones present even more disturbing scenarios. A drone carrying aerosolized biological agents or chemical weapons could release payloads over populated areas, potentially affecting thousands. The technical requirements for such attacks are not trivial, weaponizing biological or chemical agents requires expertise but they are not insurmountable for determined groups with access to scientific knowledge. Several terrorism analysts have identified this scenario as among the most likely applications of drone technology to mass casualty attacks.

Radiological dispersal, so-called "dirty bombs" represents another concerning application. While acquiring sufficient radioactive material for significant contamination poses challenges, small amounts could be dispersed via drone over politically sensitive locations, causing panic and requiring expensive decontamination even if actual health risks were minimal. The psychological and economic damage from such attacks could exceed physical effects significantly.

Assassination via drone has already occurred in multiple contexts and will likely increase. The 2018 attempted assassination of Venezuelan President Maduro using explosive-laden drones demonstrated feasibility, though execution failed. Future attempts with more sophisticated planning and better equipment will likely succeed. Political leaders, military commanders, business executives, and other high-value individuals face new vulnerabilities that traditional security measures struggle to address.

Critical infrastructure presents targets where relatively small attacks could cause disproportionate effects. Power substations, communication facilities, transportation hubs, and industrial plants often have minimal physical security because their vulnerability was limited historically. Drones change this calculus. A few well-placed explosive charges delivered by drones could cause blackouts, disrupt communications, or damage industrial processes with cascading economic effects.

The defensive challenge that drone terrorism poses is severe. Current security measures such as metal detectors, vehicle barriers and armed guards prove largely ineffective against aerial threats. Counter-drone systems exist but are expensive, require technical expertise, and often cannot be deployed in environments where civilians operate personal drones legally. The cost of defending every potential target against drone attack would be astronomical and likely impossible in open societies.

The psychological dimension of drone terrorism may prove as significant as physical effects. The knowledge that attacks could come from above at any time, that public gatherings are vulnerable, that critical infrastructure lacks protection, erodes the sense of security essential to normal social functioning. Even without frequent attacks, the threat itself imposes costs through fear, changed behaviors, and expensive security measures of questionable effectiveness.

Regional Conflicts Transformed

Beyond terrorism, drone proliferation will reshape regional conflicts globally, affecting territorial disputes, ethnic conflicts, and proxy wars in ways that alter strategic calculations and power balances. The Middle East, South Asia, East Asia, Africa, and Latin America will all see conflicts transformed by technologies demonstrated in Ukraine.

The Middle East faces particularly significant drone proliferation impacts given existing conflicts, extensive Iranian drone programs, and regional arms races. The Israeli-Palestinian conflict will see increased drone employment by both sides, with Israeli defense forces expanding counter-drone capabilities while Palestinian groups develop more sophisticated attack systems. The Yemen conflict, already featuring extensive drone use, will intensify further as Houthi forces improve capabilities and Saudi Arabia develops more effective defenses and strike systems.

Iran's position as both proliferator and direct participant in regional drone warfare will expand. Iranian proxies across the Middle East, in Syria, Lebanon, Iraq, and Yemen will receive increasingly capable systems and training. Iranian drone attacks against regional adversaries, particularly Saudi Arabia and potentially Israel, will grow in sophistication and frequency. The regional powers opposing Iran will respond with their own drone programs and counter-drone capabilities, driving rapid capability growth across the region.

South Asian conflicts, particularly the India-Pakistan rivalry and disputes in Kashmir, will incorporate lessons from Ukraine's drone war. Both nations have substantial defense industrial bases, large militaries, and ongoing conflicts where drones could provide significant advantages. The relatively short distances between population centers and military installations in South Asia make drone strikes particularly viable. An Indo-Pakistani conflict featuring mass drone employment would represent the first major conventional war between nuclear powers since drone technology matured, a scenario with terrifying escalation potential.

The China-Taiwan situation and broader U.S.-China competition will be fundamentally shaped by drone proliferation. Taiwan's defense strategy increasingly emphasizes asymmetric capabilities, recognizing that conventional force-on-force competition with China is unsustainable.

Drones provide asymmetric options both for defending against Chinese amphibious operations and for striking Chinese logistics and command facilities. The PLA's massive drone acquisition programs reflect recognition that swarm attacks could overwhelm Taiwanese defenses. The balance of power in the Taiwan Strait will be determined partly by who fields more effective drone swarms and counter-drone systems.

African conflicts, while typically lower-intensity than major wars, will be transformed by drone proliferation in ways that exacerbate instability and complicate peacekeeping. Groups operating in vast, sparsely monitored territories will use drones for reconnaissance and strike capabilities that magnify their effectiveness against government forces. Ethnic conflicts will be fought with drone surveillance enabling targeted attacks on civilian populations. Resource conflicts in mineral-rich regions will see drones employed to control territory and logistics routes.

Latin American conflicts, particularly those involving drug cartels and insurgent groups, already feature limited drone employment that will expand dramatically. Cartels use drones for smuggling across borders, for reconnaissance against law enforcement, and occasionally for targeted killings. As capabilities improve and costs decline, cartel drone operations will become more sophisticated and prevalent. Colombian and Mexican governments face adversaries with substantial financial resources that can afford cutting-edge drone technology, creating law enforcement challenges unlike anything previously encountered.

The Proliferation Challenge

International efforts to control drone proliferation face daunting obstacles that may prove insurmountable. Unlike nuclear weapons or chemical weapons, which require specialized materials and manufacturing facilities that could be monitored and restricted, drones can be built from commercial components using publicly available knowledge. Traditional non-proliferation frameworks such as export controls, inspection regimes and international treaties prove largely ineffective against technologies so widely distributed and easily assembled.

The Missile Technology Control Regime (MTCR), established to prevent proliferation of delivery systems for weapons of mass destruction, includes drones in its scope but achieves limited effectiveness. MTCR restrictions apply to complete military systems and sophisticated components, but cannot address commercial items or prevent determined actors from acquiring capabilities. Many major drone-producing nations are not MTCR participants. Even participating states struggle to enforce controls given dual-use technology prevalence.

Attempts to establish international treaties specifically governing military drones have foundered on disagreements about scope, definitions, and verification. Some nations advocate banning autonomous weapons systems entirely; others oppose restrictions that would limit their own military capabilities. Reaching consensus among diverse states with competing interests has proven impossible. The diplomatic processes that produced nuclear non-proliferation treaties and chemical weapons conventions cannot be replicated for technologies where the fundamental components have enormous legitimate commercial applications.

Export controls on complete drone systems provide minimal constraint when adversaries can build comparable systems domestically or acquire them from non-controlled sources. Chinese manufacturers, operating outside Western export control regimes, provide systems to any purchaser without restrictions. Iranian production, while less sophisticated, supplies proxies and allies throughout the Middle East. Turkish drones have been sold widely across Africa and Asia. The global supply of capable systems vastly exceeds any control mechanisms.

Domestic regulations restricting civilian drone operations, while appropriate for airspace management and privacy protection, do little to prevent malicious use. Criminals and terrorists planning attacks will not be deterred by regulations requiring drone registration or restricting flight near airports. The regulations burden legitimate users while providing minimal security benefits against determined adversaries.

The fundamental reality is that drone technology is out, widely distributed, and cannot be put back under control. The knowledge, components, and capabilities exist globally in quantities that make comprehensive restriction impossible. Future proliferation efforts must therefore focus not on preventing spread which is futile but on developing effective defenses, establishing norms around employment, and maintaining intelligence capabilities to detect and disrupt malicious planning before attacks occur.

A World Transformed

The proliferation of drone technology will transform global security environments more completely than any military technology since nuclear weapons, and arguably more so because drones will be employed far more frequently in far more contexts. The world entering 2030 will be one where aerial threats are ubiquitous, where no location is truly secure from attack, where asymmetric capabilities enable small groups to threaten large nations, and where defensive measures struggle perpetually to keep pace with offensive innovations.

For individuals, this transformed world means reduced security and increased vigilance. Large gatherings will require counter-drone protection. Critical infrastructure will need hardening against aerial attack. Public figures will require security measures accounting for drone assassination threats. The psychological burden of living in societies where threats can materialize from any direction at any time will affect quality of life and social cohesion.

For governments, the challenges are existential. Protecting populations against drone threats while preserving civil liberties and economic functionality will prove enormously difficult. The cost of comprehensive counter-drone coverage for entire nations would be prohibitive. Achieving security will require uncomfortable tradeoffs between protection and freedom, between surveillance and privacy, between defensive spending and other priorities.

For international relations, drone proliferation creates new instabilities and lowers barriers to conflict. The ability to strike adversaries with minimal risk and cost encourages aggressive behavior. Arms races in drone and counter-drone capabilities will consume resources and increase tensions. Conflicts that might have remained limited could escalate as proliferated drone capabilities enable rapid offensive operations.

The historical parallel is not to nuclear weapons, which created stability through mutual vulnerability and high employment barriers, but to machine guns and artillery in the early twentieth century, technologies that made warfare vastly more lethal and fundamentally transformed how conflicts were fought. The decades following those technologies' proliferation saw unprecedented warfare. Whether the twenty-first century follows a similar trajectory depends partly on whether humanity learns from history and partly on factors beyond any single nation's control.

Ukraine did not create drone technology, but Ukrainian innovation and Russian employment demonstrated its military potential in ways that will echo globally for decades. The proliferation now underway is a consequence of that demonstration, spread of knowledge and capabilities that cannot be contained, and transformation of warfare that every nation and non-state actor must now accommodate. The genie is well and truly out, flying on millions of drones worldwide, and the world must adapt to this new reality whether prepared or not.

Chapter 32

NATO's Adaptation

The North Atlantic Treaty Organization enters the post-Ukraine period facing a reckoning. For seven decades, NATO military doctrine, force structure, and procurement priorities centered on conventional combined arms warfare emphasizing technological superiority, professional forces, and expensive platforms designed to defeat Soviet and later Russian military power. Ukraine's war has revealed that this paradigm, while not obsolete, is incomplete and perhaps inadequate for the character of future conflict. The alliance now undertakes the most significant adaptation since its founding, attempting to absorb lessons from Ukraine while maintaining capabilities for traditional scenarios, restructuring forces for drone-centric warfare while preserving conventional strength, and doing all of this within political and budgetary constraints that make comprehensive transformation difficult.

European Defense Planning Post-Ukraine

The shock of Ukraine's experience has reverberated through European defense ministries with varying intensity. Nations closest to Russia such as Poland, the Baltic states and Romania absorbed lessons immediately and are restructuring forces rapidly. Western European nations, more distant from perceived threats and facing different political pressures, move more slowly but are moving nonetheless. The cumulative effect is a wholesale reassessment of European defense planning that will reshape military capabilities for decades.

Poland, sharing borders with both Ukraine and Belarus, understands viscerally that its security depends on capabilities demonstrated in Ukrainian fighting. Polish defense planning now centers on drone warfare as the primary operational paradigm rather than supporting capability. The Polish military is reorganizing brigade structures to integrate drone units at all levels, establishing dedicated drone battalions, and creating new command structures for managing distributed operations. Polish defense budgets, already substantial at over three percent of GDP, are being reoriented toward drone procurement and domestic production capability.

Polish domestic industry is being mobilized for drone production through government contracts, joint ventures with Ukrainian manufacturers, and incentives for private sector investment. The goal is indigenous production capacity of hundreds of thousands of drones annually, eliminating dependence on foreign suppliers and creating an industrial base for wartime surge capacity. Polish planning explicitly assumes future conflict with Russia and designs force structure for high-intensity attritional warfare where industrial capacity determines outcomes.

The Baltic states of Estonia, Latvia and Lithuania face similar imperatives with more limited resources. Their small populations and economies cannot support Ukrainian-scale production, but they can develop specialized capabilities and contribute to NATO collective defense. Estonian investment focuses on electronic warfare and counter-drone technologies, leveraging the nation's cybersecurity expertise. Latvian and Lithuanian programs emphasize integration with Polish and other allied forces, interoperability standards, and niche capabilities that complement larger allies' contributions.

Romania, holding NATO's southeastern flank along the Black Sea, has initiated comprehensive military modernization emphasizing drone capabilities and lessons from Ukraine's defense of its coastline and southern regions. Romanian planning accounts for threats from the Black Sea, potential instability in Moldova, and the need to defend extensive territory with limited forces. Drones provide Romania with force multipliers enabling credible defense despite numerical disadvantages against potential adversaries.

Western European nations such as Germany, France, the United Kingdom, Italy and Spain approach adaptation more cautiously but with greater resources. These nations maintain global commitments beyond European territorial defense, possess sophisticated defense industries with vested interests in traditional platforms, and face domestic political contexts where rapid militarization proves challenging. Yet even these nations recognize that adaptation is necessary.

Germany, after decades of military underinvestment, has pledged massive defense spending increases and is reorienting procurement toward capabilities demonstrated in Ukraine. German defense planning now includes drone acquisition programs measuring in hundreds of thousands of units, domestic production initiatives partnering German engineering expertise with Ukrainian operational experience, and comprehensive counter-drone networks for German territory and forward-deployed forces.

France, traditionally emphasizing independent military capability and sophisticated technology, is integrating lessons from Ukraine while maintaining its preference for advanced systems. French defense planning envisions high-low mix: sophisticated long-range strike drones and surveillance systems complemented by mass production of simpler tactical systems. The French defense industry sees opportunities in post-war Ukrainian reconstruction and in serving as European leader in drone technology, competing with German and other initiatives.

The United Kingdom, despite Brexit complicating European defense cooperation, remains committed to NATO and is undertaking significant adaptation. British defense planning incorporates Ukrainian lessons into concepts for territorial defense of the British Isles, forward deployment in Eastern Europe, and expeditionary operations globally. British forces are experimenting with integrated drone warfare doctrine, testing operational concepts where drones and traditional forces operate in sophisticated combined arms teams.

Massive Drone Acquisition Programs

The scale of NATO drone procurement, planned and underway, represents expenditure measuring in hundreds of billions of dollars over coming years. This investment reflects recognition that drone warfare is not a temporary phenomenon or theater-specific requirement but rather fundamental transformation requiring complete force structure adaptation.

The United States, while not European, remains NATO's dominant military power and is undertaking drone procurement at scales that dwarf European efforts. U.S. defense planning now incorporates requirements for millions of tactical drones, hundreds of thousands of long-range strike systems, and comprehensive counter-drone capabilities for deployed forces and homeland defense. The U.S. military's global commitments mean that requirements exceed even Ukrainian-scale operations, forces must be equipped for simultaneous operations across multiple theaters against adversaries who will field their own mass drone capabilities.

American drone procurement follows multiple parallel paths. Traditional defense contractors are developing sophisticated systems incorporating advanced AI, extensive range, and multiple mission profiles. These programs produce expensive, capable platforms in relatively modest quantities. Simultaneously, the Department of Defense is pursuing alternative acquisition strategies for simpler systems producible at scale, partnering with non-traditional vendors, streamlining approval processes, and accepting greater risk in pursuit of faster fielding.

The tension between these approaches reflects broader debates within the American defense establishment about how to balance quality versus quantity, sophistication versus simplicity, and traditional procurement versus rapid innovation. Some senior officers advocate for Ukrainian-style emphasis on mass production of good-enough systems; others argue that

American technological advantages should be preserved and that sophisticated systems will prove superior. The debate is not merely academic, procurement decisions made now will determine American military effectiveness in conflicts years hence.

European procurement, while smaller in absolute terms than American programs, represents proportionally similar or greater commitment given smaller defense budgets. Collectively, European NATO members plan to acquire millions of drones over the next decade, representing tens of billions of dollars in procurement spending. This expenditure requires reallocating budgets from other priorities, reducing orders for fighter jets and other expensive platforms, and making difficult tradeoffs between competing requirements.

The procurement strategies vary by nation but generally emphasize domestic production capability, interoperability with allies, and rapid acquisition timelines. Unlike traditional defense procurement measuring timelines in decades, drone programs are structured for delivery within months or a few years. This acceleration requires accepting greater risk, reduced testing, and willingness to field imperfect systems that can be improved through iteration rather than demanding perfection before deployment.

European defense industries are positioning themselves to capture procurement spending through domestic production initiatives, joint ventures, and partnerships. Major contractors like Airbus, BAE Systems, Leonardo, and Dassault are developing drone programs ranging from tactical systems to sophisticated long-range platforms. Simultaneously, smaller companies and startups are emerging, often founded by veterans of Ukraine's drone industry or by entrepreneurs recognizing market opportunities.

The collective European procurement represents industrial policy as much as military acquisition. European governments understand that drone production capability is a strategic asset essential for security independence, economic competitiveness, and technological leadership. Procurement decisions therefore consider industrial policy objectives: building domestic manufacturing, creating employment, developing technological expertise, and reducing dependence on non-European suppliers, particularly Chinese component manufacturers.

Domestic Production Initiatives

The recognition that wartime industrial capacity determines outcomes in drone warfare has sparked European initiatives to develop domestic production capability independent of foreign suppliers. This industrial mobilization represents a profound shift from peacetime procurement emphasizing international supply chains and lowest-cost sourcing toward wartime readiness prioritizing security of supply and surge capacity.

Poland's domestic production initiative exemplifies this approach. The Polish government has established public-private partnerships with Ukrainian manufacturers, licensing designs proven in combat and adapting them for Polish production. Polish facilities are being built or converted to drone manufacturing with the explicit goal of reaching annual production capacity of 200,000 units by 2027. This capacity would enable Poland to sustain Ukrainian-level attrition rates independently or contribute substantially to broader NATO operations.

Polish initiative includes complete supply chain development: electronic component manufacturing, motor and propulsion system production, battery fabrication, and final assembly. The comprehensiveness reflects determination to achieve genuine independence rather than merely assembling imported components. Polish planners understand that in major conflict, relying on Chinese electronics or other foreign inputs creates unacceptable vulnerability.

Baltic cooperation on drone production creates regional capacity exceeding what any single small nation could achieve. Estonian, Latvian, and Lithuanian governments have established joint procurement agencies and are developing specialized production facilities in each country: Estonia focusing on electronic warfare systems and counter-drone technologies, Latvia on reconnaissance systems and ground control infrastructure, Lithuania on tactical strike systems. The distributed production model provides redundancy while enabling efficient specialization.

Germany's industrial mobilization leverages existing manufacturing sophistication to develop production capacity for advanced systems. The German defense industry can already produce sophisticated drones in modest quantities; the challenge is scaling to Ukrainian levels while maintaining quality. The German approach emphasizes automation, precision manufacturing, and integrated quality control enabling rapid production without sacrificing reliability. German facilities under construction or expansion will add capacity for 150,000 drones annually by 2028.

French production initiatives balance mass production with technological sophistication, reflecting the French defense industry's traditional emphasis on advanced systems. French manufacturers are developing production lines for multiple drone categories: simple tactical systems produced in large quantities, sophisticated long-range platforms in smaller numbers, and specialized systems for particular missions. French capacity will reach approximately 100,000 units annually across all categories.

United Kingdom domestic production focuses on systems optimized for British military requirements while maintaining export potential. British manufacturers are partnering with Ukrainian companies to access combat-proven designs while contributing British engineering expertise and manufacturing capability. UK production capacity goals are modest by continental standards, approximately 50,000 units annually reflecting smaller military size and emphasis on quality over sheer quantity.

The collective European production initiative aims for continental capacity exceeding one million drones annually within five years. This capacity would enable European NATO to sustain high-intensity operations independently of American support, a strategic autonomy goal that has gained urgency following Ukraine's experience and concerns about American political reliability. European production would also position Europe as a major drone exporter, competing with Chinese, Turkish, Israeli, and American manufacturers in global markets.

Doctrine and Training Overhaul

Military hardware alone does not create capability; forces must understand how to employ new systems effectively. NATO is undertaking comprehensive doctrinal revision and training transformation to integrate drone warfare into all aspects of military operations. This represents a cultural and intellectual challenge as significant as the material and industrial adaptation.

Traditional NATO doctrine, developed over decades and validated through exercises and limited conflicts, emphasized sequential operations: gaining air superiority, achieving information dominance, enabling ground maneuver through precision fires. Drone warfare disrupts this sequence. When adversaries can attack throughout one's depth using cheap, numerous systems, air superiority becomes impossible to achieve absolutely. When both sides possess extensive reconnaissance capabilities, information dominance becomes contested continuously. When precision fires are available to lowest tactical levels, ground maneuver occurs under constant observation and threat.

New doctrine under development addresses these changed conditions by assuming continuous contested operations across all domains. Rather than phases of conflict where one side achieves superiority enabling subsequent operations, NATO doctrine envisions persistent multi-domain combat where advantages are temporary and local. Success comes through better adapting to dynamic conditions, faster OODA loop execution, and superior integration of capabilities rather than through achieving comprehensive dominance.

Tactical doctrine is being rewritten to account for drone proliferation. Small unit tactics now incorporate dispersal to reduce vulnerability to drone strikes, emphasis on camouflage and concealment against aerial observation, and organic counter-drone capabilities at squad and platoon levels. Infantry sections integrate drone operators as standard positions. Vehicle crews train to evade drone attacks and employ countermeasures. Every soldier receives basic training in detecting and reporting drones.

Operational doctrine addresses force employment in environments saturated with drones. Planning factors account for attrition rates orders of magnitude higher than historical norms. Logistics planning assumes supply lines under constant drone attack requiring defended convoys and hardened facilities. Maneuver planning incorporates drone operations as primary rather than supporting capability, with ground forces enabling drone employment as much as drones support ground maneuver.

Training transformation is occurring at all levels from individual soldiers to theater commanders. Basic training now includes drone awareness, identification of friendly and enemy systems, and individual protective measures. Advanced training for specialists covers drone operation, maintenance, and repair. Unit training exercises integrate drone operations into all scenarios, with opposing forces employing Ukrainian-style mass drone tactics to prepare NATO forces for realistic threats.

NATO's Joint Warfare Centre in Norway has established a comprehensive drone warfare training program drawing on Ukrainian advisors and lessons learned. The program hosts officers from all NATO nations, providing immersive training in planning and conducting drone-centric operations. Exercises simulate Ukrainian-intensity combat with hundreds of drones employed by both sides, creating realistic training that exposes students to operational demands they will face in future conflicts.

Professional military education is being revised to incorporate drone warfare throughout curricula. Staff colleges are rewriting programs to address planning and logistics in high-attrition environments. Command courses emphasize decision-making under conditions of distributed operations where commanders cannot directly observe battles. War colleges examine strategic implications of drone proliferation and industrial warfare's return.

The cultural adaptation required should not be underestimated. NATO militaries have spent decades developing professional armed forces built around sophisticated platforms and highly trained personnel. Drone warfare emphasizes different attributes: rapid adaptation, distributed initiative, acceptable failure rates, and industrial sustainability. Officers and NCOs whose careers centered on traditional paradigms must adapt to radically different conditions or make way for younger personnel more comfortable with new approaches.

The Ukrainian Model vs. Traditional Military Procurement

Perhaps the most difficult adaptation for NATO involves reconciling Ukrainian innovation's speed and pragmatism with Western military procurement's deliberate risk-averse processes. The contrast could not be starker: Ukrainian garage workshops producing combat-ready drones within weeks versus NATO programs requiring years from requirement definition to fielding. This tension reflects different institutional contexts, legal frameworks, and cultural assumptions about acceptable risk and appropriate oversight.

Ukrainian procurement success reflects wartime conditions that eliminate peacetime constraints. Environmental regulations, worker safety requirements, quality certification processes, and oversight mechanisms that would slow or prevent rapid production are suspended or ignored. Financial controls are relaxed to enable immediate funding. Competition and market mechanisms replace bureaucratic procurement procedures. Failure is accepted as inevitable rather than treated as a scandal requiring investigation.

NATO cannot and arguably should not replicate these conditions in peacetime. Democratic accountability requires oversight preventing corruption and waste. Safety regulations protect workers and civilians. Quality standards ensure that systems function reliably. Competitive procurement ensures fairness and efficiency. These are not merely bureaucratic obstacles but reflections of values that distinguish democratic societies from authoritarian alternatives.

Yet NATO also recognizes that excessively rigid procurement processes create vulnerabilities. When adversaries can innovate and field new systems faster than NATO can respond, procedural correctness becomes strategic liability. The challenge is determining how much risk and how much procedural flexibility is acceptable while maintaining essential oversight and accountability.

Various NATO nations are experimenting with alternative acquisition approaches attempting to capture Ukrainian agility within democratic frameworks. "Rapid acquisition authorities" bypass standard processes for urgent requirements, accepting greater risk in exchange for speed. "Other Transaction Authorities" enable contracting with non-traditional vendors using commercial rather than government contracting procedures.

Prototyping and experimentation programs field limited quantities for testing before full production decisions.

These alternative approaches show promise but face limits. They work for small programs acquiring modest quantities but scale poorly to mass production measuring in hundreds of thousands or millions. They create oversight challenges when traditional accountability mechanisms are bypassed. They can be exploited for corruption or waste without careful management. Most fundamentally, they exist as exceptions to standard processes rather than as standard processes themselves.

The deeper question is whether NATO procurement can be reformed comprehensively rather than merely creating exceptions around standard procedures. This would require legislative changes, cultural transformation, and accepting substantially different risk tolerances. Some nations, particularly those closest to Russian threats, are attempting such reforms. Others maintain traditional approaches, trusting that peacetime thoroughness will prove superior to wartime improvisation over the long term.

Ukraine provides empirical evidence that speed and adaptation matter enormously, that good-enough systems fielded quickly outperform perfect systems delayed indefinitely, and that distributed innovation can match or exceed centralized planning. Whether NATO can learn these lessons while maintaining democratic oversight and social values remains to be seen. The alliance's adaptation to drone warfare will be judged not merely by how many systems are procured but by whether procurement processes themselves are transformed to enable the sustained innovation that future conflicts will demand.

Preparing for the Next War

NATO's adaptation ultimately aims not to refight Ukraine's war but to prepare for the next war, which will feature even more drones, greater autonomy, more sophisticated threats, and adversaries who have also learned from Ukrainian experience. The alliance must therefore look beyond current capabilities toward future requirements, invest in technologies not yet mature, and develop concepts for operations not yet imaginable in detail.

Future NATO planning assumes adversaries will field drone capabilities at least matching and potentially exceeding alliance capabilities. Russia will rebuild military power incorporating lessons from Ukraine. China will leverage its industrial advantages to produce drones at scales NATO cannot match. Regional powers will acquire sophisticated systems from multiple suppliers. NATO planning therefore cannot assume technological superiority or numerical advantage but must prepare for conflicts where adversaries possess comparable or superior drone capabilities.

Operational concepts under development address combat in highly contested airspace where thousands of drones operate simultaneously. These concepts emphasize decentralization, with small units capable of independent operations when communications are jammed or denied. They incorporate extensive use of AI and autonomy, recognizing that human operators cannot manage the complexity and speed of future drone combat. They assume high attrition requiring constant replacement and accepting temporary capability gaps.

NATO's long-term vision extends beyond current technological capabilities toward emerging possibilities. Hypersonic drones, directed energy weapons, AI-enabled swarms, and other advanced systems will reshape warfare further. Alliance planning must somehow balance investment in current-generation capabilities needed immediately against developing next-generation systems required for future dominance. This balancing act will define NATO military effectiveness for decades.

The adaptation NATO undertakes now will determine whether the alliance remains relevant and effective military organization or becomes an obsolete relic of an earlier era. The stakes could not be higher. Ukraine has provided warning and lessons; whether NATO heeds them adequately will be tested in future conflicts that may come sooner than any would wish. The drones are coming, indeed they are already here and NATO's survival as an effective defensive alliance depends on adapting to the reality they represent.

Chapter 33

The Lessons We Can't Ignore

Ukraine's drone war, when viewed comprehensively across its trajectory from improvised desperation in 2022 to industrial-scale operations in 2025, offers lessons of profound and uncomfortable clarity. These are not subtle insights requiring sophisticated analysis to discern but rather stark, undeniable truths that challenge fundamental assumptions about modern warfare, military effectiveness, and strategic planning. The lessons are uncomfortable because they contradict decades of Western military thinking, undermine justifications for vast procurement expenditures, and suggest that the future of warfare looks less like precision-guided high-tech combat and more like industrialized attrition reminiscent of conflicts thought relegated to history. These lessons cannot be ignored without accepting potentially catastrophic vulnerability in future conflicts.

Attrition Warfare Returns

The first and most fundamental lesson is that attrition warfare—the grinding contest of materiel consumption and replacement that characterized World War I, World War II, and Korea has returned as the dominant paradigm for high-intensity conflict. For decades, Western military planning assumed that precision weapons, information dominance, and technological superiority would enable swift, decisive campaigns where objectives were achieved before attrition became a significant factor. The Gulf War, Kosovo, and initial phases of Iraq and Afghanistan operations seemed to validate this assumption.

Ukraine demolishes it entirely. Despite sophisticated reconnaissance, precision strikes, and advanced weapons, neither Russia nor Ukraine has achieved swift decisions. Instead, the conflict has devolved into sustained attritional struggle where daily losses measure in thousands of drones, hundreds of armored vehicles, and casualties that would shock publics accustomed to low-casualty interventions. The side that can replace losses faster ultimately prevails, regardless of tactical excellence or technological sophistication.

The implications challenge core assumptions of modern military planning. If attrition dominates, then sustainability matters more than initial capability. The force that can maintain operations for months or years proves superior to one optimized for short intensive campaigns. This favors different force structures, procurement priorities, and industrial policies than have guided Western militaries for decades.

Attrition warfare at scale requires accepting casualties and material losses that contemporary Western societies find politically intolerable. Ukrainian forces sustain operations despite attrition rates that would trigger political crises in Western democracies. Russian willingness to accept enormous casualties to achieve marginal gains reflects different political calculus than democratic societies can maintain. This asymmetry creates strategic vulnerability: adversaries willing to accept higher costs can exploit Western sensitivity to casualties.

The return of attrition warfare also means that conflicts will be longer than recent Western experience suggests. The expectation that modern wars will be swift derives from limited interventions against inferior adversaries. Peer or near-peer conflicts will not follow this pattern. Ukraine and Russia have fought for over three years with no decisive conclusion despite enormous commitments by both sides. Future major wars may last years or even decades, requiring sustained industrial production, population mobilization, and political will that few Western nations are prepared to maintain.

Industrial Capacity Trumps Technological Edge

The second inescapable lesson is that industrial capacity, the ability to produce and replace military systems at scale determines outcomes more than technological sophistication of individual platforms. This lesson directly contradicts Western procurement philosophy emphasizing expensive, highly capable systems produced in modest quantities.

Ukraine and Russia both learned this lesson through bitter experience. Early in the conflict, sophisticated Turkish Bayraktar drones achieved spectacular successes but could not be replaced rapidly enough. Ukrainian forces sometimes operated with only a few Bayraktars across entire fronts. Conversely, simple FPV drones assembled from commercial components could be produced in thousands monthly and replaced immediately when destroyed. The numerous simple systems proved strategically decisive while expensive sophisticated platforms became footnotes.

Russia's experience reinforces the lesson. Sophisticated Russian cruise missiles and hypersonic systems grab headlines but can be produced only in limited numbers. Shahed drones, simple and unsophisticated, can be manufactured by thousands monthly using smuggled Western electronics and Chinese components. Russia's ability to sustain nighttime terror campaigns against Ukrainian cities depends entirely on Shahed production volumes, not on sophisticated systems costing millions.

This reality challenges fundamental assumptions underlying Western defense procurement. NATO militaries invest vast sums in platforms like F-35 fighters, costing over $100 million per aircraft and requiring years to produce. A single F-35 costs equivalent to 200,000 FPV drones. The F-35 is incomparably more capable in many respects, but can it deliver 200,000 times the military effect? When drones can be lost freely because replacements arrive constantly, while F-35 losses would be catastrophic, the cost-benefit calculation shifts dramatically.

The industrial capacity lesson extends beyond production to maintenance, logistics, and sustainability. Simple systems that can be repaired in the field with basic tools and common parts prove more sustainable than sophisticated platforms requiring specialized facilities and components. Ukrainian workshops repair damaged drones using salvaged parts and improvised techniques. Western systems often require returning to depots or even manufacturers for repairs, creating dependencies that become critical vulnerabilities in high-intensity conflict.

Mobilization capacity, the ability to surge production rapidly in response to war, represents a critical component of industrial capacity that Western nations have largely abandoned. Ukrainian drone manufacturers scaled from zero to millions annually in three years. Chinese industry could plausibly scale to tens of millions annually if mobilized. Western defense contractors, operating with peacetime business models and specialized facilities, lack comparable surge capacity. This asymmetry creates strategic vulnerability that adversaries could exploit.

The Myth of Surgical Precision

Western military doctrine for decades has emphasized precision strikes that minimize collateral damage, distinguish carefully between military and civilian targets, and achieve strategic effects through carefully calibrated force. This doctrine rests on assumptions that Ukraine's experience reveals as questionable at best and dangerous at worst.

Precision is expensive. GPS-guided munitions, laser-guided bombs, and cruise missiles cost tens or hundreds of thousands of dollars each. Their expense creates scarcity that limits employment. During peacetime interventions against inferior adversaries with limited air defenses, this scarcity proves acceptable because few munitions are needed. In high-intensity peer conflict, scarcity becomes a crippling constraint. When hundreds of targets require engagement daily and expensive precision munitions are exhausted rapidly, what then?

Ukraine's answer is volume over precision. FPV drones lack sophisticated guidance but compensate through human operators and numbers. Miss rates are higher than precision munitions achieve, but cost allows multiple attempts. The cumulative effect of thousands of relatively imprecise strikes proves more militarily significant than dozens of precise ones. This reality contradicts decades of doctrine emphasizing quality over quantity.

The surgical strike ideal also assumes conflicts where minimizing civilian casualties and collateral damage is paramount. This assumption holds for Western expeditionary operations where political considerations constrain military action heavily. It fails in existential conflicts where both sides prioritize military effectiveness over civilian protection. Russia demonstrates willingness to conduct indiscriminate attacks causing mass civilian casualties. Ukraine, while attempting to follow humanitarian law, must conduct operations in populated areas where civilian harm is often unavoidable. The surgical precision ideal becomes a luxury that combatants in existential struggles cannot afford.

Western militaries built around precision strike doctrine face difficult adaptation. Their force structures, procurement priorities, and operational concepts assume precision will be available and sufficient. When precision proves unaffordable or insufficient in high-intensity conflict, these militaries may lack mass-strike capabilities necessary for effectiveness. The F-35s, advanced destroyers, and precision missile inventories that dominate Western arsenals cannot be rapidly converted to mass-production simple systems that attritional warfare demands.

Cheap and Numerous Beats Expensive and Few

The corollary to industrial capacity's importance is that cheap numerous systems defeat expensive scarce ones through arithmetic that no amount of sophistication can overcome. This lesson manifests repeatedly in Ukraine but contradicts fundamental assumptions underlying Western military investment.

The cost asymmetry between drone attack and defense illustrates the principle starkly. A $500 FPV drone can destroy a $5 million air defense missile or a $50 million tank. Even if air defenses intercept 90% of attacking drones, the economics remain unfavorable. Russia can afford to launch ten drones knowing nine will be intercepted if the tenth destroys a valuable target. Ukraine cannot afford to expend air defense missiles at rates required to intercept all drones, the defender will exhaust expensive interceptors before the attacker exhausts cheap drones.

This asymmetry extends across military systems. Cheap naval drones threaten expensive warships. Simple reconnaissance drones locate targets for artillery costing fraction of precision missiles. Mass-produced FPV systems destroy armored vehicles costing millions. In each case, the cheap numerous system imposes costs on expensive scarce systems that cannot be sustained.

The temptation is to argue that sophisticated counter-systems will restore balance, but this misunderstands the fundamental mathematics. Developing expensive counter-drone systems merely creates new expensive targets for cheap drones to attack. The defender always faces economic disadvantage because defense must protect everything while offense can choose targets. When attacking systems cost orders of magnitude less than defending systems or protected assets, the arithmetic favors attackers decisively.

NATO militaries structured around expensive platforms face existential challenges from this reality. The alliance has invested trillions in sophisticated systems produced in limited quantities: fighter jets, main battle tanks, warships, strategic bombers. These systems face threats from cheap drones costing tiny fractions of their value. NATO cannot realistically defend all these expensive assets against sustained drone attack while simultaneously maintaining offensive capabilities. Something must give, either force structure must change radically or NATO must accept vulnerability that adversaries will certainly exploit.

The lesson extends to personnel as well. Professional militaries composed of highly trained, expensive soldiers face asymmetric costs when confronting adversaries fielding large numbers of minimally trained operators. Training an F-35 pilot costs millions and takes years. Training an FPV drone operator costs thousands and takes weeks. The professional soldier represents much greater investment that creates asymmetric risk, losing elite personnel is catastrophic while adversaries absorb similar losses readily.

Innovation in Combat vs. Peacetime Procurement

Ukraine demonstrates that innovation accelerates dramatically under combat pressure in ways that peacetime procurement cannot replicate. The gap between Ukrainian innovation speed and Western procurement timelines reveals an uncomfortable truth: the systems and processes that Western militaries built to ensure oversight, prevent waste, and maintain accountability also prevent the rapid adaptation that warfare demands.

Ukrainian manufacturers iterate drone designs weekly based on operator feedback. Problems identified in Monday morning's combat inform Tuesday afternoon's design modifications implemented by Thursday. This cycle repeats continuously, producing systems that evolve in real-time with tactical requirements. Western procurement, by contrast, measures timelines in years or decades from requirement definition through development, testing, approval, and production. Systems fielded may address requirements that are years obsolete.

The Ukrainian advantage stems from eliminating bureaucratic overhead, accepting risk that Western systems prevent, and prioritizing speed over perfection. Quality control is minimal, systems that function adequately are fielded immediately rather than waiting for exhaustive testing. Safety regulations are ignored or minimized. Environmental compliance is disregarded. Competitive procurement is bypassed through direct negotiation. All of this violates principles that Western democracies insist upon but proves remarkably effective at generating combat capability rapidly.

The question Western militaries face is whether peacetime procurement can be reformed to capture wartime innovation advantages without abandoning essential oversight. The track record suggests pessimism is warranted. Attempts at acquisition reform occur regularly but achieve limited success because the fundamental tension between speed and accountability cannot be resolved easily. Democratic societies demand transparency, competition, and oversight that inherently slow processes. Eliminating these safeguards risks corruption and waste that undermine public support for defense spending.

Some argue that Western militaries should maintain two parallel procurement systems: traditional processes for major platforms requiring careful development, and rapid acquisition pathways for simpler systems where speed matters more than perfection. This approach has merit but creates coordination challenges, risks resource competition between systems, and may simply formalize different categories of slow procurement rather than enabling genuine speed.

The deeper lesson may be that peacetime military preparation inevitably differs from wartime requirements and that accepting this gap is necessary. Western militaries will continue procuring sophisticated systems through deliberate processes while hoping that wartime exigency will trigger innovation acceleration when needed. This approach risks being caught unprepared if future conflicts erupt without a warning period for adaptation, but may represent realistic accommodation to institutional realities.

Preparing for Mass Drone Warfare

The final inescapable lesson is that future conflicts will feature drone warfare at scales that current preparations barely begin to address. Ukraine and Russia together will have produced and employed perhaps five to seven million drones by the time their conflict concludes. This represents merely the beginning of drone proliferation, not its culmination. Future conflicts, particularly those involving major powers, could see tens or hundreds of millions of drones employed over their duration.

NATO militaries, despite increasing drone acquisitions, are not preparing for warfare at these scales. Current plans envision hundreds of thousands or perhaps low millions of drones, substantial increases over current inventories but insufficient for sustained high-intensity operations. The industrial capacity being developed can support current planning but would prove inadequate if conflicts exceed assumed intensity or duration.

The gap between current preparations and actual requirements will become evident only when tested in combat, at which point closing it may be impossible. Industrial mobilization requires years to achieve scale. Training operators demand time that combat will not provide. Developing doctrine and tactics cannot be rushed beyond certain limits. The nation or alliance that enters future conflict better prepared for mass drone warfare will possess decisive advantages that may prove impossible to overcome.

Preparing adequately requires accepting uncomfortable implications. Defense budgets must shift dramatically from expensive platforms to mass production of simpler systems. Force structures must be reorganized around drone operations as primary capability. Industrial policy must prioritize defense production capacity. Military culture must adapt to accepting higher attrition and loss rates. Political leaders must explain to the public that warfare's character has changed in ways requiring different approaches.

These adaptations face resistance from institutional inertia, defense industry lobbying, political constraints, and genuine uncertainty about how radically to break from established practices. The temptation is to hedge, making some adaptations while maintaining traditional capabilities and structures but hedging may prove insufficient when conflicts test preparations.

The Hard Truth

The lessons from Ukraine are clear, uncomfortable, and undeniable. Attrition warfare has returned. Industrial capacity trumps technological sophistication. Precision is a luxury that high-intensity conflict cannot afford. Cheap numerous systems beat expensive scarce ones through mathematics that cannot be overcome. Peacetime innovation cannot match wartime urgency. Future conflicts will feature drone warfare at scales that mock current preparations.

These lessons contradict decades of Western military planning, undermine vast procurement investments, and demand adaptations that institutions struggle to make. They can be rationalized, minimized, or explained away as theater-specific or adversary-specific phenomena not applicable to NATO. This would be a comfortable but potentially catastrophic mistake.

The hard truth is that warfare has fundamentally changed in ways that favor different capabilities, structures, and approaches than Western militaries have prioritized. The nation or alliance that adapts most completely to this new reality will possess decisive advantages in future conflicts. Those that fail to adapt, that cling to comfortable assumptions or resist difficult changes, will find themselves at devastating disadvantage when tested.

Ukraine has provided warnings, examples, and lessons. Whether they are heeded adequately will be determined in future conflicts that may come sooner than anyone wishes. The drones have transformed warfare permanently. The only question is whether militaries worldwide will transform themselves quickly enough to remain effective in the age of mass drone warfare that Ukraine's experience has inaugurated.

Epilogue

The War That Changed Everything

On a cold February morning in 2022, Russian forces crossed into Ukraine expecting swift victory. Their columns advanced confidently toward Kyiv, their generals anticipated capitulation within days, and their president prepared victory speeches. The world watched, assuming that Ukraine would fall as quickly as Afghanistan had to the Taliban months earlier. The conventional wisdom held that modern warfare between such asymmetric forces would produce swift, decisive outcomes favoring the larger military power.

Three and a half years later, the world has learned otherwise. Ukraine still fights. Russia still struggles. The swift decisive victory never materialized, replaced instead by grinding attritional warfare that has consumed hundreds of thousands of lives, exhausted vast arsenals, and transformed military technology more completely than any conflict since World War II. The transformation happened not through the deployment of secret weapons or revolutionary technologies but through the mass proliferation of simple, cheap, effective systems that have redefined what modern warfare means.

The Transformation 2022-2025

In February 2022, Ukraine possessed virtually no indigenous drone manufacturing. By October 2025, the country produces 2.5 million drones annually with capacity expanding toward 4 million. In early 2022, drones were supporting assets used sparingly for reconnaissance and occasional strikes. By late 2025, drones are primary combat systems employed in millions, with entire operational concepts built around their capabilities. In 2022, a single Turkish Bayraktar destroying a Russian column made international headlines. By 2025, dozens of drone strikes will occur every hour along hundreds of kilometers of front, generating so much footage that individual incidents barely register.

The numbers themselves tell the story of transformation's scale. Russia has launched over 5,600 Shahed drones and nearly 200 missiles through September 2024 alone, with campaigns intensifying thereafter. Ukraine's production reached 1.5 million FPV drones in 2024. Individual manufacturers like TAF Drones produce 40,000 units monthly valued at over $1 billion annually. The October 9, 2025 attack saw over 400 Russian drones launched in a single night. These figures, incomprehensible at the conflict's beginning, have become routine, markers of how completely drones have come to dominate operations.

The technological evolution has been equally dramatic. Early systems were commercial quadcopters and racing drones pressed into military service with minimal modification. By 2025, Ukraine and Russia field sophisticated dedicated military platforms: long-range strike drones reaching hundreds of kilometers deep into enemy territory, fiber-optic guided FPV systems immune to jamming, AI-enabled reconnaissance platforms with autonomous target recognition, and interceptor drones designed specifically for counter-drone missions. The progression from improvisation to industrialization occurred in timeframes that traditional military development cycles measure in decades.

The tactical transformation extends beyond technology to doctrine, organization, and culture. Ukrainian forces operate with decentralization and initiative that would have been impossible in Soviet-legacy military structures. Operators trained in weeks conduct complex missions independently. Volunteer networks provide capabilities that formal military procurement cannot match. Garage workshops produce systems that compete with establishment defense contractors. The entire ecosystem of warfare has been remade around drone operations in ways that challenge every assumption about how militaries function.

The Permanent Change to Warfare

What Ukraine has demonstrated is not temporary phenomenon or theater-specific adaptation but permanent transformation in warfare's character. The technologies, tactics, and industrial processes proven in Ukrainian skies will spread globally and define military competition for decades. Future conflicts, regardless of geographic location or participating parties, will feature drone warfare at scales that make Ukraine's experience seem like a preliminary demonstration rather than a culmination.

The permanence derives from fundamental economics and physics that cannot be reversed. Drones are cheap to produce using commercial components and publicly available designs. They are effective against targets ranging from individual soldiers to strategic infrastructure. They are proliferating globally through channels that cannot be effectively controlled. These characteristics ensure that drone warfare will expand rather than contract, intensify rather than diminish, and spread rather than remain contained.

The strategic implications are profound and irrevocable. The concept of strategic depth, safety through distance from front lines has collapsed. Rear areas, once secure by virtue of being beyond enemy reach, are now vulnerable to attacks from cheap platforms capable of flying hundreds of kilometers. Cities far from combat zones wake nightly to air raid sirens as drones penetrate deep into national territory. Critical infrastructure, economic centers, and civilian populations face constant threat regardless of geographic position.

The military implications challenge every assumption underlying modern force structure and procurement. Expensive sophisticated platforms that dominated twentieth century warfare face obsolescence or radical role revision when cheap drones can destroy them for tiny fractions of their cost. Air superiority, the foundation of Western military doctrine, becomes impossible to achieve absolutely when adversaries can field thousands of drones daily. Precision strike, long emphasized as a hallmark of advanced military capability, proves unaffordable in conflicts where thousands of targets require engagement.

The political implications extend beyond military operations to questions of deterrence, arms control, and international stability. When warfare becomes affordable and accessible, the barriers to conflict lower. When attacks can be conducted remotely without risking soldiers, domestic political constraints on military action weaken. When non-state actors can acquire capabilities previously requiring nation-state resources, the monopoly on organized violence that defines statehood erodes. The world emerging from Ukraine's experience will be more dangerous, more unstable, and more violent than the one that existed before.

Ukraine's Sacrifice and Innovation

The permanent transformation of warfare has been purchased with Ukrainian blood, treasure, and suffering on scales that defy adequate acknowledgment. Hundreds of thousands dead or wounded. Millions displaced from homes. Cities reduced to rubble. Infrastructure systematically destroyed. Three and a half years of constant attack, sleepless nights, and existential threat. Ukraine paid these costs not by choice but by necessity, the price of national survival against genocidal aggression.

Yet within this horror, Ukraine has achieved innovation that will shape global military competition for generations. The nation transformed from having no indigenous drone capacity to becoming world leader in operational expertise, manufacturing capability, and tactical innovation. Ukrainian engineers, operators, and entrepreneurs have solved problems that defense establishments in wealthy nations struggled to address. They have demonstrated production models that combine agility with scale. They have proven that rapid innovation under pressure can outpace peacetime development despite vastly fewer resources.

Ukraine's innovation emerged not from research laboratories or carefully planned programs but from desperate necessity, creative improvisation, and willingness to accept risks that more comfortable societies avoid. Garage workshops replaced formal factories. Volunteer networks substituted for government procurement. Operator feedback drove design iteration rather than bureaucratic requirements. The results have been systems that work, tactics that succeed, and capabilities that established military powers now study intensively and attempt to replicate.

The knowledge Ukraine has built through three and a half years of continuous combat represents an asset of immeasurable value. No amount of peacetime exercise or simulation can replicate learning that comes from actual warfare. Ukrainian drone operators possess expertise that no other military can claim. Ukrainian manufacturers understand production challenges and solutions that competitors can only theorize about. Ukrainian commanders have developed doctrine written not in classrooms but in blood and verified through combat results rather than theoretical analysis.

This knowledge will be Ukraine's contribution to the world long after the war ends. Nations worldwide will study Ukrainian experience, adapt Ukrainian innovations, and apply Ukrainian lessons to their own military preparations. Ukraine has become an inadvertent teacher to global militaries, demonstrating possibilities and revealing vulnerabilities that will inform defense planning for decades. The terrible price Ukraine paid purchases an understanding that might spare others from similar suffering, small consolation for Ukrainians but real contribution to humanity nonetheless.

The Drone Age Is Here

The fundamental reality is that warfare has entered a new age characterized by ubiquitous aerial threats, mass production of lethal autonomous systems, and combat conducted increasingly through remote platforms rather than direct human engagement. This drone age is not approaching or emerging, it has arrived, fully formed and accelerating rapidly toward even more complete dominance.

Evidence surrounds us. Ukraine produces millions of drones annually and plans to expand further. Russia manufactures comparable quantities. China's industrial capacity enables production at scales that dwarf both combatants if mobilized. NATO members collectively plan acquisitions measuring in millions of units. Regional powers worldwide are establishing drone production facilities and acquiring capabilities. Non-state actors are incorporating drones into operational concepts. The proliferation is global, rapid, and irreversible.

The technologies enabling drone warfare continue improving exponentially. Artificial intelligence advances enable greater autonomy. Battery technology extends range and endurance. Manufacturing processes reduce costs. Sensor capabilities enhance effectiveness. Communication systems increase resilience. Each advancement makes drones more capable, more accessible, and more central to military operations. The trajectory points toward a future where human operators become supervisors rather than pilots, where swarms coordinate autonomously, and where aerial combat occurs at speeds and scales beyond human comprehension.

The industrial mobilization Ukraine demonstrated establishes a template that other nations will follow when faced with existential threats. The shift from peacetime production to wartime output at scales measuring in millions will be replicated globally as conflicts emerge. The garage workshops, volunteer networks, and distributed production models that seemed uniquely Ukrainian innovations will prove universally applicable to nations with technical expertise and manufacturing infrastructure. The industrialization of warfare that Ukraine pioneered will become a standard feature of major conflicts.

No Going Back

Perhaps the most important recognition is that there is no returning to pre-2022 conditions. The knowledge exists, the technologies are proliferated, the industrial processes are established, and the tactical innovations are spreading globally. No treaty, export control regime, or international agreement can stuff the genie back into the bottle. The transformation is permanent and irreversible.

This permanence imposes obligations on nations, militaries, and societies worldwide. Ignoring the lessons from Ukraine, hoping that conflicts will remain limited, or assuming that traditional approaches remain adequate is not merely complacent but potentially catastrophic. The nation or alliance caught unprepared for mass drone warfare will face devastating disadvantages in future conflicts. Preparation is not optional but rather an existential requirement.

For military organizations, preparation means comprehensive restructuring around drone warfare as the primary operational paradigm. Force structures must integrate drones at all levels. Procurement must shift toward mass production of affordable systems. Doctrine must account for combat in environments saturated with aerial threats. Training must prepare personnel for warfare's transformed character. Industrial policy must ensure capacity for sustained production at wartime scales.

For governments, preparation requires difficult political choices about resource allocation, industrial policy, and strategic priorities. Defense spending must increase substantially or be reallocated dramatically from traditional platforms to new capabilities. Regulations must be reformed to enable rapid innovation while maintaining essential oversight. International cooperation must be strengthened to share costs and combine capabilities. The population must be prepared psychologically for warfare's changed character and its potential costs.

For societies, preparation involves reckoning with uncomfortable realities about vulnerability, threat, and the price of security. The civilian sanctuary that international humanitarian law attempted to protect has been eroded by technologies that reach anywhere. The distinction between front lines and rear areas that provided some populations with safety has collapsed. The expectation that modern warfare would be swift and decisive has been replaced by recognition that future major conflicts may be long, attritional, and demanding of sustained sacrifice.

What We Must Remember

As years pass and as Ukraine's war eventually concludes, distance and time will tempt forgetting or minimizing what has been learned. New crises will command attention. Different threats will seem more pressing. The specifics of Ukrainian experience will fade from memory. This forgetting would be a dangerous mistake.

We must remember that cheap, numerous systems defeated expensive, sophisticated ones through mathematics that sophistication cannot overcome. We must remember that industrial capacity determines outcomes more than tactical excellence or technological advantage. We must remember that innovation under pressure outpaced peacetime development despite vastly fewer resources. We must remember that attrition warfare returned despite decades of assumptions that precision and technology had made it obsolete.

We must remember Ukraine's resilience, creativity, and determination in face of existential threat. We must remember the volunteer networks that crowdfunded millions of drones. We must remember the garage workshops that outperformed defense contractors. We must remember the operators who fought from relative safety yet bore psychological burdens that traditional combat never imposed. We must remember the civilians who lived under constant drone attack, who adapted to conditions that should have been intolerable, and who maintained dignity and humanity despite everything.

Most fundamentally, we must remember that warfare has changed permanently and that preparation for future conflicts requires accepting uncomfortable truths that challenge assumptions and demand difficult adaptations. The drone age is here. The transformation is permanent. The lessons are clear. Whether we remember them adequately and act upon them decisively will determine outcomes of conflicts yet to come and potentially the survival of nations and alliances in the transformed strategic environment that Ukraine's war has created.

The drones are flying. They will continue flying in ever-greater numbers and sophistication. The world must adapt to this reality or face consequences that Ukraine's terrible experience has so clearly illuminated. This is the lesson that cannot be forgotten, the truth that cannot be ignored, and the challenge that defines our historical moment. Ukraine has provided a warning. Whether we heed it is a decision that every nation, every military, and every society must make and live with the consequences of that choice.

www.ingramcontent.com/pod-product-compliance
Ingram Content Group UK Ltd.
Pitfield, Milton Keynes, MK11 3LW, UK
UKHW020154290926
15471UKWH00005B/870

9 798232 474355